Springer Theses

Recognizing Outstanding Ph.D. Research

Aims and Scope

The series "Springer Theses" brings together a selection of the very best Ph.D. theses from around the world and across the physical sciences. Nominated and endorsed by two recognized specialists, each published volume has been selected for its scientific excellence and the high impact of its contents for the pertinent field of research. For greater accessibility to non-specialists, the published versions include an extended introduction, as well as a foreword by the student's supervisor explaining the special relevance of the work for the field. As a whole, the series will provide a valuable resource both for newcomers to the research fields described, and for other scientists seeking detailed background information on special questions. Finally, it provides an accredited documentation of the valuable contributions made by today's younger generation of scientists.

Theses are accepted into the series by invited nomination only and must fulfill all of the following criteria

- They must be written in good English.
- The topic should fall within the confines of Chemistry, Physics, Earth Sciences, Engineering and related interdisciplinary fields such as Materials, Nanoscience, Chemical Engineering, Complex Systems and Biophysics.
- The work reported in the thesis must represent a significant scientific advance.
- If the thesis includes previously published material, permission to reproduce this must be gained from the respective copyright holder.
- They must have been examined and passed during the 12 months prior to nomination.
- Each thesis should include a foreword by the supervisor outlining the significance of its content.
- The theses should have a clearly defined structure including an introduction accessible to scientists not expert in that particular field.

More information about this series at http://www.springer.com/series/8790

Christopher J. Ballance

High-Fidelity Quantum Logic in Ca⁺

Doctoral Thesis accepted by
the University of Oxford, Oxford, UK

 Springer

Author
Dr. Christopher J. Ballance
Department of Physics
University of Oxford
Oxford
UK

Supervisor
Dr. David Lucas
Department of Physics
University of Oxford
Oxford
UK

ISSN 2190-5053   ISSN 2190-5061   (electronic)
Springer Theses
ISBN 978-3-319-88563-6   ISBN 978-3-319-68216-7   (eBook)
DOI 10.1007/978-3-319-68216-7

Printed on acid-free paper

This Springer imprint is published by Springer Nature
The registered company is Springer International Publishing AG
The registered company address is: Gewerbestrasse 11, 6330 Cham, Switzerland

Supervisor's Foreword

The field of quantum information processing dates back at least thirty years, to the ideas of Richard Feynman and David Deutsch. They pointed out that, while quantum mechanical systems involving more than a few particles are extremely difficult to simulate on conventional computers, because the number of variables required to describe the quantum system increases exponentially with the number of particles, it ought to be possible to take advantage of this same exponential growth to perform computations in ways not accessible to classical computers. About twenty years ago, interest in quantum computing was greatly stimulated by Peter Shor's discovery of an algorithm for a quantum computer which could factorize large numbers, a notoriously difficult problem for classical computers. Indeed, our systems of public key cryptography (used, e.g., to protect transactions on the Internet) rely on the practical impossibility of factorization of sufficiently large integers.

However, for many years, it seemed that such computational power would never be more than a theorist's pipe dream: there was no obvious way to correct the inevitable errors which would corrupt the quantum state of any real-world system, because to check for errors by measuring the quantum bits, or qubits, would collapse their wavefunction and prematurely terminate the calculation in progress. The situation changed dramatically with the discovery by Shor and Robert Calderbank, and independently by Andrew Steane, of methods of quantum error correction, which could correct errors using extra qubits and logic operations, even if those qubits and operations were themselves imperfect. Unfortunately, to implement these techniques successfully requires high precision in the logic operations; despite nearly two decades of effort in optimizing methods of quantum error correction, it is not known how to build a quantum computer with logic precision below a "threshold" level of approximately 99%. Generally, the most demanding operation is a multi-qubit logic gate, which takes two or more individual qubits and entangles them into a joint quantum state. The first two-qubit gate in any physical system was demonstrated in 2000, by Chris Monroe and colleagues, working in the group of David Wineland at NIST Boulder, by laser manipulation of two trapped atomic ions, the internal atomic energy levels of which served as the qubit states; however, its precision was well below the level required for error correction methods.

A landmark experiment in 2008 performed by Jan Benhelm and colleagues, in Rainer Blatt's group at the University of Innsbruck, demonstrated a two-qubit logic gate with 99.3% fidelity, also using trapped-ion qubits. For precision only just exceeding the threshold level, though, the overhead (in terms of the number of extra qubits and logic operations) required to build even a quantum computer with 100 logical qubits is daunting: perhaps millions of physical qubits would be required, and such a machine would be prohibitively expensive to construct. This overhead falls rapidly with increasing precision, so it was imperative to improve the fidelity further. However, it was not until the experiments reported in Chris Ballance's D. Phil. thesis that a significant further advance was made, with the achievement of a two-qubit gate fidelity at the 99.9% level. He also made the first systematic study of the trade-off between speed and gate precision for trapped-ion quantum logic, in the process achieving the fastest-ever gate in a trapped-ion system. Most importantly, Ballance made a detailed study of the different sources of imperfection which account quantitatively for the observed 0.1% gate error, and showed that his error model accounted well for the measured error over more than an order of magnitude range of gate speeds. An encouraging result of this study is that all the dominant sources of error are technical in nature; that is, there are good prospects for reducing them by technical improvements to the apparatus and it ought to be possible to reduce the error by at least another order of magnitude before reaching any fundamental limitations of the atomic system he uses (qubits stored in hyperfine states of the ground level of the calcium-43 ion).

One should not get the impression that the work of building a quantum computer is now merely an engineering challenge. Further progress in improving the fidelity is required before error correction overheads become practical, and new ideas will be needed to circumvent the fundamental limitations of this particular gate method which will likely be approached in the coming few years.

In a second experiment, Ballance describes the application of the same laser-driven gate operation to entangle two different species of ion (in this case, two different isotopes of calcium). This is likely to be a powerful technique in the future: the differing optical frequencies to which the two species respond can be used to protect "memory" qubits from interacting with laser beams driving operations in other "logic" qubits; furthermore, one can take advantage of the distinct properties of different species for qubits fulfilling separate roles, for example using one species for logic and another as an "interface" qubit to link separate ion traps using single-photon interconnections—the basis of a "quantum Internet".

The results of the principal experiments described in the thesis were published in two papers in *Nature* (2015) and *Physical Review Letters* (2016), in each case alongside complementary work by the NIST group, who achieved comparable results using beryllium and magnesium trapped-ion qubits.

For experts in the field, the thesis makes reading as gripping as any thriller: it briskly summarizes the necessary theoretical background and presents the methods and results with a precision of writing (almost!) as high as that of the experiments it describes. It demonstrates a combination of mastery of the relevant theory with an intimate understanding of the apparatus, which is the hallmark of all first-rate experimental physics. The only criticism that might be made is that, in its very clarity, it makes the years of painstaking detective work in the laboratory, of tracking down those details which mattered, and of pursuing the inevitable wild goose chases that did not, all seem a little too easy. The reader should not believe it was.

Oxford, UK   Dr. David Lucas
August 2017

Abstract

Trapped atomic ions are one of the most promising systems for building a quantum computer—all of the fundamental operations needed to build a quantum computer have been demonstrated in such systems. The challenge now is to understand and reduce the operation errors to below the "fault-tolerant threshold" (the level below which quantum error correction works) and to scale up the current few-qubit experiments to many qubits. This thesis describes experimental work concentrated primarily on the first of these challenges. We demonstrate high-fidelity single-qubit and two-qubit (entangling) gates with errors at or below the fault-tolerant threshold. We also implement an entangling gate between two different species of ions, a tool which may be useful for certain scalable architectures.

We study the speed/fidelity trade-off for a two-qubit phase gate implemented in $^{43}\text{Ca}^+$ hyperfine trapped-ion qubits. We develop an error model which describes the fundamental and technical imperfections/limitations that contribute to the measured gate error. We characterize and minimize various error sources contributing to the measured fidelity, allowing us to account for errors due to the single-qubit operations and state readout (each at the 0.1% level), and to identify the leading sources of error in the two-qubit entangling operation. We achieve gate fidelities ranging between 97.1(2)% (for a gate time $t_g = 3.8$ μs) and 99.9(1)% (for $t_g = 100$ μs), representing, respectively, the fastest and lowest error two-qubit gates reported between trapped-ion qubits by nearly an order of magnitude in each case. We also characterize single-qubit gates with average errors below 10^{-4} per operation, over an order of magnitude better than previously achieved with laser-driven operations.

Additionally, we present work on a mixed-species entangling gate. We entangle a single $^{40}\text{Ca}^+$ ion and a single $^{43}\text{Ca}^+$ ion with a fidelity of 99.8(5)% and perform full tomography of the resulting entangled state. We describe how this mixed-species gate mechanism could be used to entangle $^{43}\text{Ca}^+$ and $^{88}\text{Sr}^+$, a promising combination of ions for future experiments.

Acknowledgements

Dr. David Lucas has proved an excellent supervisor—as well as sharing his impressive knowledge of atomic physics (and, for better or for worse, Turbo-Pascal and DOS), he also let me get on with whatever I wanted to do in the laboratory, signed orders, and provided much-needed guidance and critique.

Over the past few years, most of my time has been spent working with Dr. Thomas Harty, Vera Schäfer and Dr. Norbert Linke. Tom has been a springboard for many an idea (both good and bad) and has been an exceptional quartermaster. Vera has been a great addition to the "old laboratory", overcompensating for her smaller stature with phenomenal torque. I have enjoyed Norbert's company during our rebuilding of the "old laboratory", despite our differing opinions on a certain micro-fabricated trap.

Thank you also goes to Sarah Woodrow and Keshav Thirumalai, who are busy building the next generation of the experiment, and to the rest of the trapped-ion gang: Martin Sepiol, Jamie Tarlton, Diana Prado Lopes Aude Craik, and Hugo Janacek, as well as Prof. Andrew Steane and Prof. Derek Stacey, the other pillars of the group. I would also like to thank Dr. David Allcock for the ARSES.

I am forever grateful to my parents for raising a child who, from a young age, would take apart everything he could get his hands on (but took a while longer to master the art of putting them back together without any pieces left over). Your care, patience and support have enabled me to become the keen experimentalist I am today—I hope this scientific endeavour will prove less contentious than my previous ventures. My brothers have both encouraged and endured my scientific interests, listening to (and sometimes participating in) many a long technical discussion. I am grateful also for the encouragement and care of my Aunt, Shirley.

I would like to thank Isabel, who has been a constant and unswerving source of distraction over the course of my D.Phil.—without her, my life would be far less interesting.

I gratefully acknowledge funding by the EPSRC and by Hertford College's Carreras senior scholarship. I would also like to acknowledge the funding I received throughout my undergraduate studies from the Stuart Barber bursary.

Contents

Chapter 1
Introduction

The 20th century has been profoundly influenced by the development of fast, automated, information processing. One of the ground-breaking concepts that made many of the daunting theoretical and practical problems of such information processing tractable is this: the details of the physical device used for a computation do not matter – all 'useful' computers, no matter how they physically operate, are equivalent in terms of the classes of problems they can solve.

This means that rather than having to develop new computational techniques to solve the same problem on different types of physical device, we can develop techniques using a mathematically convenient model (such as a Turing Machine), and expect the results to hold for all physical devices.

Despite the power of such abstractions, we need to remember that any real computing device is built out of components that obey the laws of nature. The classical physics that computational theory relies on is only an approximation to the physical reality that quantum mechanics describes – nature is quantum, not classical. Following this line of thought Deutsch showed that a Quantum Turing Machine could efficiently solve problems that were not efficiently solvable on a classical Turning Machine [Deu85]. (Here 'efficiently' means that the difficulty scales only polynomially with problem size, rather than super-polynomially.) Such a device, that processes information at the quantum level rather than at the classical level, is known as a Quantum Computer.

1.1 The Quantum Computer

In principle we know how to build a quantum computer. First we need a system that maintains its quantum behaviour over time, that is, it is sufficiently isolated that the environment does not couple in an uncontrolled way to our computer (causing decoherence). We then need to be able to initialise our system to a known state, perform unitary operations spanning the computer's Hilbert space, and perform projective measurements on some or all of the Hilbert space. These requirements on any

© Springer International Publishing AG 2017
C.J. Ballance, *High-Fidelity Quantum Logic in Ca+*,
Springer Theses, DOI 10.1007/978-3-319-68216-7_1

candidate quantum computer, codified by DiVincenzo [DiV00], are experimentally daunting, but do not appear to be impossible.

In an extension of the classical terminology, the two-level building block of the computational space is called a quantum bit (qubit). There are an infinite number of unitary operators we may wish to apply to our system, but it has been shown that we can efficiently approximate any unitary operator from repeated application of a small set of operations on single qubits, and one two-qubit operation [Kit97].

Noise and decoherence at some level are unavoidable in any system. Over the course of executing an algorithm on a quantum computer, such errors destroy the fragile superposition that carries the algorithm's result, rendering the computation useless. However, just as in classical computers, it is possible to correct for these errors at the cost of additional resources.

The question then is what quantity of extra resources do we need to implement error correction for a given operation error-rate? First of all, our operations need to have errors below the 'fault-tolerant' threshold – this is the operation error above which the error correction algorithm fails, no matter how many additional resources are available. Some of the best error-correction algorithms proposed, such as the 'surface code', have error thresholds at the 1% level [FMMC12]. To perform a useful computation the operation error needs to be substantially below the threshold. For example, if we wanted to use a computer that had operation errors of 0.1% to factor a 2000 bit number into its prime factors using Shor's algorithm, along with surface-code error-correction, we would need 10^8 qubits [FMMC12].

Many different physical implementations for quantum computers have been proposed, including photons, quantum dots, neutral atoms, solid-state spins, superconducting Josephson junctions, and trapped ions. Some experimental results have been demonstrated for all these systems, but most have not demonstrated convincing techniques for scaling or high-fidelity operations. Currently trapped ions are the most mature implementation [MK13, BW08]; however superconducting Josephson junctions [DS13] have improved dramatically in recent years, and show great promise.

1.2 Quantum Computation with Trapped Ions

In the trapped-ion architecture, two states of an ionized atom form the qubit, with a single qubit per ion. Atoms are ideal qubits for the same reason they make good clocks: they are guaranteed to be identical the universe over, and atomic physics is superbly well tested and understood. Ionized atoms are used (over neutral atoms) as their electric charge allows for strong confinement (trap depths of millions times the Doppler limit of laser-cooling). We confine the ions, under vacuum, in traps formed by electric fields – this leaves them very well isolated from the environment, and hence sources of decoherence.

Two types of ion qubit are used; 'ground state' qubits consist of two states in the ground level of the ion, whereas 'optical' qubits use one state in the ground level and one in a metastable excited state. Optical qubits offer some advantages, but have a finite lifetime due to decay of the excited state (typically ~ 1 s). Ground state qubits, however, do not decay (lifetimes of millions of years) – coherence times of 10 minutes have been measured [BH91], and there is no fundamental limit to how far this can be increased. In this thesis we use only ground state qubits. Single-qubit and two-qubit operations are performed using optical or microwave fields, with the two-qubit operations mediated by the coupled motion of ions in a single harmonic trap. The qubit states are initialised by optical pumping, and read out using state-selective electron shelving and fluorescence.

So far we have described the building blocks. The natural way to use these blocks, and the first described [CZ95], is to confine a large number of ions in a single trap like beads on a string. By focussing laser beams single qubits, or pairs of qubits, can be manipulated in isolation. The problem with this approach is one of spectral crowding – as we add more ions it becomes harder to resolve individual motional modes (needed for the entangling gates), requiring a reduction in the operation speed. This scheme is thus limited to tens of ions, far short of the number needed.

A more promising approach is the 'Quantum CCD' architecture [WMI+98]. Here a complex set of electrodes forms a large number of interconnected traps. The qubit ions can be shuttled between different traps, and reordered in an arbitrary fashion, by varying the electrode voltages. Different regions of the QCCD are used for qubit readout, qubit storage, and entangling operations. Any arbitrary pair of qubits can be entangled by shuttling them into the same region and performing an entangling gate. This architecture is much more scalable than the 'beads on a string' approach, but it still has daunting problems; namely the density of laser and electronic access, and scheduling and management of the ion shuttling. These technical issues mean that scaling the Quantum CCD architecture to many thousands of qubits is likely to be very challenging.

The solutions to these technical problems is the 'Networked Cell' architecture [MK13, NFB14]. In this scheme one interconnects many 'small' cells. Each cell is a QCCD containing at as many qubits as can reliably be made to work (likely 5–50). Entanglement is created between cells using a photonic link. Two-dimensional nearest-neighbour links between the cells are sufficient to implement the surface code [FMMC12]. Only the intra-cell operations need to be performed with high fidelity, a relatively lossy and noisy inter-cell link is tolerable [NFB14]. The power of this scheme is that it is freely scalable. Once we can build a reliable unit cell the scalability problems change to cost (and engineering) problems.

1.3 Thesis Outline

The remainder of this thesis is structured as follows:

Chapter 2 gives an overview of the calcium ion qubit – how we trap, cool, and manipulate single ions. We then describe the quantisation of the motion of a crystal of trapped ions, and how they couple to a general travelling-wave radiation field.

Chapter 3 reviews the use of two-photon Raman transitions to manipulate ground-level qubits. We derive the basic coupling, and describe the many potential sources of error in such a coupling. We then calculate these coupling coefficients for our calcium ion qubits, and present experimental work confirming these models.

Chapter 4 gives an overview of the implementation of two-qubit gates in trapped-ion systems. It explains the concept behind the 'geometric phase' gate that is commonly used. It then discusses our specific implementation of this gate mechanism, the 'light-shift' gate. The chapter concludes with an analysis of the different sources of error in the light-shift two-qubit gate mechanism.

Chapters 5 and 6 describes the apparatus built up over the course of my D.Phil. Chapter 5 focusses on the design and construction, while Chap. 6 describes the characterization and performance of the apparatus.

Chapter 7 describe the implementation and randomized benchmarking of high-fidelity single-qubit gates implemented on the ^{43}Ca$^+$ low-field 'clock' qubit.

Chapter 8 presents the results of our two-qubit gate experiments, including a detailed analysis of the experimental errors. We also present an implementation of an entangling gate between two different isotopes of calcium ions, and describe how this could be used to entangle calcium ion qubits with strontium ion qubits.

Chapter 9 concludes this thesis, summarising our results and their relevance to the field.

References

[Deu85]   Deutsch, D. 1985. Quantum Theory, the Church-Turing Principle and the Universal Quantum Computer. *Proceedings of the Royal Society A: Mathematical, Physical and Engineering Sciences* 400 (1818): 97–117.

[DiV00]   DiVincenzo, D.P. 2000. The physical implementation of quantum computation. *Fortschritte der Physik* 48: 771–783.

[Kit97]   Kitaev, A.Y. 1997. Quantum computations: algorithms and error correction. *Russian Mathematical Surveys* 52 (6): 1191–1249.

[FMMC12]   Fowler, A.G., M. Mariantoni, J.M. Martinis, and A.N. Cleland. 2012. Surface codes: Towards practical large-scale quantum computation. *Physical Review A* 86 (3): 032324.

[MK13]   Monroe, C., and J. Kim. 2013. Scaling the Ion Trap Quantum Processor. *Science* 339 (6124): 1164–1169.

[BW08]   Blatt, R., and D.J. Wineland. 2008. Entangled states of trapped atomic ions. *Nature* 453 (7198): 1008–15.

[DS13]   Devoret, M.H., and R.J. Schoelkopf. 2013. Superconducting circuits for quantum information: an outlook. *Science* 339 (6124): 1169–74.

[BH91]   Bollinger, J.J., and D. Heizen. 1991. A 303-MHz frequency standard based on trapped Be+ ions. *IEEE Transactions on Instrumentation and Measurement* 40 (2): 126–128.

[CZ95]   Cirac, J.I. and P. Zoller. Quantum Computations with Cold Trapped Ions. *Physical Review Letters*, 74(20), 1995.

[WMI+98]   Wineland, D.J., C. Monroe, W.M. Itano, D. Leibfried, B.E. King, and D.M. Meekhof. Experimental Issues in Coherent Quantum-State Manipulation of Trapped Atomic Ions. *Journal Of Research Of The National Institute Of Standards And Technology*, 103(3), 1998.

[NFB14]   Nickerson, N.H., J.F. Fitzsimons, and S.C. Benjamin. Freely Scalable Quantum Technologies using Cells of 5-to-50 Qubits with Very Lossy and Noisy Photonic Links. *ArXiv*, June 2014.

Chapter 2
Trapped-Ion Qubits

In this chapter we give an overview of how one can use a calcium ion as a qubit. We then review the operation of a Paul trap, and discuss the quantised behaviour of a 'crystal' of trapped ions near their motional ground state. The shared motional degrees of freedom of such a crystal allow the implementation of quantum 'logic gates', i.e. multi-qubit entangling operations.

2.1 The Calcium Ion as a Qubit

In this work we use two different isotopes of singly-ionised calcium as qubits, $^{43}\text{Ca}^+$ (nuclear spin $I = 7/2$) and $^{40}\text{Ca}^+$ ($I = 0$). In the following we describe the atomic structure of the calcium ion, how we create and Doppler-cool the ions, and how we initialise, manipulate, and read out the quantum state of the ion.

2.1.1 Photo-Ionisation

Before we do anything we have to create calcium ions. This can be done by brute force – bombarding a calcium atom with energetic electrons until a valence electron is knocked off the atom – but this can ionize any species effusing from the calcium source (a resistively heated oven). Instead, we use a much more elegant scheme that is not just atomic species selective, but isotope selective (Fig. 2.1). This scheme is two-step photo-ionisation [LRH+04, GRB+01].

In this scheme we use two lasers. The first laser is resonant with the $4s^2$-$4s4p$ transition in neutral calcium (423 nm, line-width ~ 35 MHz). As the isotope shifts in this transition are ~ 1 GHz (larger than the line-width, provided that a Doppler-free geometry is used) only one isotope is excited to the $4s4p$ level. The second laser

© Springer International Publishing AG 2017
C.J. Ballance, *High-Fidelity Quantum Logic in Ca+*,
Springer Theses, DOI 10.1007/978-3-319-68216-7_2

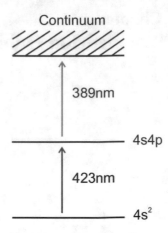

Fig. 2.1 Energy level diagram for neutral calcium showing the transitions relevant to the photo-ionisation process

(389 nm) provides enough energy to excite from the 4s4p level to the continuum, but not enough to reach the continuum from the ground state. This isotope selectivity is very useful, as it means we can selectively load one of the several isotopes of calcium in our oven, simply by adjusting the frequency of the 423 nm laser.

2.1.2   Doppler Cooling

To detect the presence of the ions, and to cool their motion, we scatter photons off the ions, and detect the scattered photons with an imaging system. To scatter photons we excite the $4S_{1/2}-4P_{1/2}$ (397 nm) transition with a laser. From this near-cycling transition there is a $\sim 5\%$ probability of decay to the $3D_{3/2}$ state (Fig. 2.2). We repump this lost population by driving the $3D_{3/2}-4P_{1/2}$ (866 nm) transition.

In an isotope without nuclear spin (e.g. $^{40}Ca^+$) this can be efficiently done with monochromatic radiation. In $^{43}Ca^+$ ($I = 7/2$) this requires two frequencies of 397 nm light to repump the ≈ 3.2 GHz $S_{1/2}$ hyperfine splitting, and is somewhat less efficient [Szw09, Har13, Jan14].

2.1.3   The Qubit and Coherent Manipulations

We want the two qubit states ($|0\rangle$ and $|1\rangle$) to be as similar in nature as possible, so that environmental perturbations do not cause a differential energy shift of the qubit (leading to decoherence). We also need to be able to prepare into one of the qubit states to initialise the qubit, and be able to read out the qubit state precisely in a single-shot.

The qubits we use in this thesis all consist of two states in the ground level $4S_{1/2}$. These states do not spontaneously decay (if we prepare an atom in one of the ground states, it will stay in that state indefinitely), but do lose their phase coherence due to magnetic field noise. In $^{40}Ca^+$ there are only two ground states, so we do not have a

Fig. 2.2 Energy level diagram for the *low* lying levels of Ca$^+$. The wavelengths and branching ratios of the dipole-allowed transitions are shown. The hyperfine splitting in ^{43}Ca$^+$ is shown for the $4S_{1/2}$ level only. The precise transition frequencies and decay rates are given in Appendix A

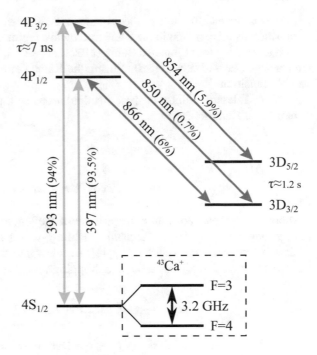

Fig. 2.3 The $4S_{1/2}$ ground level of ^{43}Ca$^+$, with the two qubits we use labelled

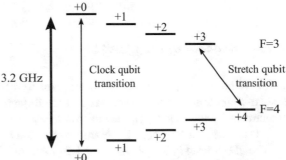

choice of qubit. In ^{43}Ca$^+$ there are 16 ground states, leading to many potential qubits (Fig. 2.3). The two we commonly use are the 'stretch' qubit ($S_{1/2}^{4,+4} - S_{1/2}^{3,+3}$) and the 'low-field clock' qubit ($S_{1/2}^{4,0} - S_{1/2}^{3,0}$).

The stretch qubit can be prepared and read out with high fidelity by optical pumping, but is highly magnetically sensitive (2.45 kHz/mG), whereas the clock qubit is harder to prepare and read out, but much less magnetically sensitive (4.8 Hz/mG for $B_0 = 2G$).[1]

---

[1]The clock qubit is first order insensitive to magnetic field variation at zero magnetic field, but we need to apply a quantisation field of $\sim 2G$ in order to read out and prepare the qubit efficiently, giving rise to a small first-order dependence.

We manipulate the state of these qubits in two ways. We directly drive the qubit transition using radiation in the RF to microwave regime, or we drive the qubit via a third level using the stimulated Raman effect. As we see in Chap. 3 there are several advantages (and disadvantages) to using the Raman effect over directly driving the qubit transition.

We can describe either of these manipulations as a pseudo-spin coupling. Our interaction Hamiltonian is

$$H_I = -\boldsymbol{\mu} \cdot \boldsymbol{B} \tag{2.1}$$
$$\boldsymbol{\mu} = \mu_m \boldsymbol{S}$$
$$\boldsymbol{B} = B\hat{\boldsymbol{x}} \cos{(\boldsymbol{k}.\boldsymbol{r} - \omega t + \phi)}$$

where $\boldsymbol{\mu}$ is our pseudo-spin, $\boldsymbol{r}$ the position of the ion, and $\boldsymbol{k}, \omega, \phi$ the wave-vector, frequency, and phase of the applied (travelling-wave) radiation. The pseudo-spin basis is our two qubit states $\{|{\uparrow}\rangle, |{\downarrow}\rangle\}$. Moving into the interaction picture with respect to the spin ($H_0 = \frac{1}{2}\omega_0\sigma_z$) and dropping far off-resonant terms,

$$H_I = \frac{\Omega}{2}\sigma_+ \exp{(i\boldsymbol{k}.\boldsymbol{r} - i\delta t + i\phi)} + \text{h.c.} \tag{2.2}$$

where $\delta := \omega - \omega_0$, and $\Omega = -\mu_m B$ is our Rabi frequency.

2.1.4 Readout of the Qubit State

After we have manipulated the qubit we want to read-out the result, i.e. detect the state of a single atom – this is generally a hard thing to do. However the technique of 'electron shelving' [Deh75] allows us to amplify this small difference in atomic state into a very large difference in fluorescence rate, and hence detect the state accurately.

The $3D_{5/2}$ level is long-lived ($\tau \approx 1.2$ s) and outside the Doppler cooling cycle ($4S_{1/2}-4P_{1/2}-3D_{3/2}$): if the ion is in $3D_{5/2}$ it will not scatter any 397 nm photons. By counting the photons scattered by the ion we can quickly detect, with high fidelity, if the ion is in the $4S_{1/2}$ state or the $3D_{5/2}$ state [MSW+08, Bur10].

We now need a way of mapping one of our $4S_{1/2}$ qubit states to $3D_{5/2}$ while not affecting the other. The technique we use for $^{40}\text{Ca}^+$ involves using a weak σ^+ polarised 393 nm beam and an intense σ^- polarised 850 nm beam [MSW+04]. With these beams there is a two-photon resonance for one of the qubit states that suppresses population transfer out of $4S_{1/2}$ while the other qubit state is freely pumped out of $4S_{1/2}$ to $3D_{5/2}$ ($\sim 90\%$) and $3D_{3/2}$ ($\sim 10\%$). The favourable branching ratios into the $3D_{5/2}$ shelf leads to a maximum shelving fidelity of 90%, hence a minimum *average* readout error of $\bar{\epsilon} = \frac{1}{2}(\epsilon_D + \epsilon_S) = 5\%$.

In $^{43}\text{Ca}^+$ we implement a simpler and more robust scheme [MSW+08, Szw09]. We assume that one of the states we want to readout out is $4S_{1/2}^{4,+4}$, and that the other is $4S_{1/2}^{3,*}$. A σ^+ polarised 393 nm beam tuned to the $4S_{1/2}^4 \leftrightarrow 4P_{3/2}^5$ transition (22 MHz linewidth) excites the population in $4S_{1/2}^{4,+4}$ to $4P_{3/2}^{5,+5}$, but does not excite population out of $S_{1/2}^3$ due to the $\sim$3 GHz hyperfine splitting. The state $4P_{3/2}^{5,+5}$ can only decay to $3D_{5/2}$ (as desired), back to the qubit state the population started in ($4S_{1/2}^{4,+4}$), or to three well-defined states in $3D_{3/2}$. If we apply appropriately polarised repumping pulses on the 850 nm transition we can recover all the population out of $3D_{3/2}$, accurately mapping one of the qubit states to $3D_{5/2}$ while leaving the other qubit state in $4S_{1/2}$ with ultimate (theoretical) errors of $\approx 1 \times 10^{-4}$.

2.2 Linear Paul Traps

For the work in this thesis we want to confine a number of ions in a 3d harmonic trap – we don't care too much about the details of this confinement. A linear Paul trap performs this job admirably.

No static electric field arrangement can spatially confine in all three dimensions. If we expand the electric field about the centre of the trap and drop all but the first order terms we have

$$\frac{F}{q} = E = \alpha x + \beta y + \gamma z \tag{2.3}$$

Poisson's equation ($\nabla \cdot E = 0$) then dictates that $\alpha + \beta + \gamma = 0$, and thus that a static arrangement of electric fields can confine in at most two dimensions, and is anti-confining in the remaining direction(s).

A linear Paul trap generates its confining potential with an oscillating (RF) 2D quadrupole field in the radial plane and with a static quadrupole electric field in the axial dimension (z). The radial fields are generated by two pairs of 'blades' and the axial by a pair of 'end-caps'. In our apparatus one pair of these blades is connected to ground, and the other to an RF source with (zero-peak) voltage V_{RF}. The electric potentials generated by the RF voltage on the blades and the DC voltage on the endcaps are described, for small displacements from the trap centre, by

$$U_{RF} = Q_x^{RF} x^2 - Q_y^{RF} y^2 + Q_z^{RF} z^2 \qquad U_{DC} = Q_z^{DC}[z^2 - \frac{1}{2}(x^2 + y^2)]$$

$$Q_x^{RF} = \frac{\alpha_x V_{RF}}{\rho_0^2} \qquad\qquad Q_y^{RF} = \frac{\alpha_y V_{RF}}{\rho_0^2}$$

$$Q_z^{RF} = \frac{\alpha_z V_{RF}}{2z_0^2} \qquad\qquad Q_z^{DC} = \frac{\alpha_z V_z}{z_0^2} \tag{2.4}$$

where ρ_0 and z_0 are the blade-centre and endcap-centre spacings. The dimensionless parameters α characterise the geometry or 'efficiency' of the electrodes (hyperbolic infinitely-long RF electrodes have $\alpha_{x,y} = 1$).

Charged particles in the trap will feel, averaged over a period of the RF, a harmonic radial force. A derivation of this ponderomotive potential is outlined in [LL76]. When the Mathieu q parameter $q \ll 1$ the motion of a particle in the trap can be approximated as being the sum of secular motion and (small amplitude) micro-motion at the RF frequency. Substituting the trap potentials (Eq. 2.4) into the equation of motion and using the standard Mathieu equation approximations we find

$$q = -\frac{4e}{m\Omega^2} \begin{pmatrix} Q_x^{RF} \\ -Q_y^{RF} \\ Q_z^{RF} \end{pmatrix} \qquad a = \frac{8e}{m\Omega^2} Q_z^{DC} \begin{pmatrix} -\frac{1}{2} \\ -\frac{1}{2} \\ 1 \end{pmatrix} \qquad (2.5)$$

$$r_i \approx r_0 \cos{(\omega_i t)} \left(1 + \frac{q_i}{2} \cos{\Omega t}\right) \qquad \omega_i = \frac{\Omega}{2}\sqrt{\frac{q_i^2}{2} + a_i} \qquad (2.6)$$

where r_i is the approximate solution for small q, and ω_i is the secular frequency of the motion.

In general there is an axial RF potential gradient. This is due to the asymmetric driving of the blades. This can be understood by first considering a symmetrically driven trap, with one blade pair driven by RF with amplitude $+V_{RF}/2$, the other blade pair driven by RF with an amplitude $-V_{RF}/2$, and the end-caps at RF ground. In this case there is a line of zero potential along the axis of the trap. We now add an (oscillating) offset of $V_{RF}/2$ to all the electrodes. The blades are now at V_{RF} and 0, and the end-caps at $V_{RF}/2$. There is still no axial RF potential gradient. If we now hold the end-caps at RF ground we break the symmetry, and there is an axial RF potential gradient: the potential at the centre of the trap oscillates, but the potential from the end-caps is static. Thus for an asymmetrically driven trap there is an axial RF pseudo-potential, and more importantly, axial micro-motion.

2.2.1 Micro-Motion Detection and Compensation

Consider a static electric field displacing the ion from the RF potential null. The ion now sees an oscillating force, leading to periodic motion of the ion at the trap RF frequency – this is micro-motion. Excess micro-motion can also be caused by an RF phase difference between the driven trap blades, which leads to the RF null moving over the RF period [BMB+98]. In both cases the resulting micro-motion leads to Doppler broadening of or sidebands on the optical transitions, which can be undesirable.

We can detect micro-motion by correlating the ion fluorescence with the trap RF phase: the varying instantaneous velocity over the motional period gives rise to a varying Doppler shift. The amplitude of the correlation reveals the amplitude of the micro-motion projected onto the beam direction. By adding static electric fields to minimise the micro-motion seen in three linearly independent beam directions we can ensure the ion sits as close as possible to the RF potential null.

2.3   Motion of the Ions

In this section we discuss the quantized motion of a crystal of ions and introduce our notation.

We assume the (identical) ions are confined in a 3d harmonic trap with radial confinement much tighter than the axial. When sufficiently cold, the ions then form linear crystals along the axis. We define the weak axis as $\hat{z}$, and the two principle radial directions as $\hat{x}$, $\hat{y}$. The motion of the N ions can then be described by N normal modes in each of these three principal directions.

2.3.1   Quantization of Motion

We wish to find an expression for r_n, the displacement of ion n from its equilibrium position. We do this by quantizing each of the normal modes of the motion, giving

$$r_n = \sum_{\hat{q}=\hat{x},\hat{y},\hat{z}} \sum_{j=1}^{N} \hat{q}\, b_n^{(\hat{q},j)} \tilde{q}_{\hat{q},j} \left(a_{\hat{q},j} + a_{\hat{q},j}^{\dagger} \right) \qquad (2.7)$$

where the first sum is over the principal axes, and the second is over the N normal modes in each direction. The normal mode matrix $b_n^{(\hat{q},j)}$ describes the amplitude of motion of ion n for the j'th normal mode in the direction $\hat{q}$, normalised such that the $\sum_n \left(b_n^{(\hat{q},j)} \right)^2 = 1$ [Jam98]. The ground state wave-function size for each normal mode is given by $\tilde{q}_{\hat{q},j} = \sqrt{\hbar/(2m\omega_{\hat{q},j})}$.

For a single ion there is only one motional mode in each direction, the centre of mass (CoM) modes. For two ions the axial modes are the centre of mass mode $b^{(\hat{z},1)} = \frac{1}{\sqrt{2}}(1,1)$ and the breathing mode $b^{(\hat{z},2)} = \frac{1}{\sqrt{2}}(1,-1)$. The radial modes are the centre of mass modes and the rocking modes (with the same normal mode coordinates).

2.3.2   Coupling to the Motion

For a general coupling between a spin and a travelling wave field (Eq. 2.2) the spatial dependence of the field leads to a coupling between spin and motion – we now consider the effect of this coupling. For simplicity we assume we only couple to the axial modes ($k \propto \hat{z}$) and we write $b_n^{(\hat{z},j)} = b_n^{(j)}$. We quantize the motion and find the motion phase for ion n to be

$$\boldsymbol{k} \cdot \boldsymbol{r}_n = \boldsymbol{k} \cdot \hat{z} \sum_j b_n^{(j)} \tilde{q}_j \left(a_j + a_j^\dagger \right) \tag{2.8}$$

$$= \sum_j \eta_n^{(j)} \left(a_j + a_j^\dagger \right) \tag{2.9}$$

where the Lamb–Dicke parameter, defined as

$$\eta_n^{(j)} = \boldsymbol{k} \cdot \hat{z} \tilde{q}_j b_n^{(j)} \tag{2.10}$$

gives the coupling strength of the applied field to the motion of ion n and mode j. This depends on the angle between the coupling wave-vector and the mode direction ($\boldsymbol{k} \cdot \hat{z}$), the motional mode frequency (via the spatial extent of the ground state wave packet, $\tilde{q}$), and the amplitude of motion of the ion for this particular mode ($b_n^{(j)}$).

We can now explicitly rewrite the motion phase factor from Eq. 2.2 in the interaction picture of $H_0 = \sum_j \omega_j (a_j^\dagger a_j + \frac{1}{2})$

$$\exp\left(i\boldsymbol{k} \cdot \boldsymbol{r}_m\right) =$$

$$\sum_{\{n'\}\{n\}} |\{n'\}\rangle\langle\{n'\}| \prod_{j=1}^{N} \exp\left(i\eta_m^{(j)} \left(a_j + a_j^\dagger\right)\right) e^{i\omega_j(n_j' - n_j)t} |\{n\}\rangle\langle\{n\}| \tag{2.11}$$

where $\{n\}$ represents all of the combinations of $n_1, n_2, ..., n_N$.

For one ion ($N = 1$) in the Lamb–Dicke regime ($\eta\sqrt{1 + 2n} \ll 1$) we can expand the coupling in a small angle approximation

$$\mathcal{H}_I = \frac{\Omega}{2} \sigma_+ \left(1 + i\eta(a^\dagger e^{i\omega_z t} + a e^{-i\omega_z t})\right) \exp\left(-i\delta t + i\phi\right) + \text{h.c.} \tag{2.12}$$

We can see that, assuming $\Omega \ll \omega_z$, we can tune to address the carrier ($\delta = 0$) and flip the spin without changing the motional state, or we can tune to the sidebands ($\delta = \pm\omega_z$) and add or subtract motional quanta while flipping the spin. In this limit the carrier Rabi frequency, Ω, is independent of motional state, and the sideband Rabi frequency is $\eta\Omega\sqrt{n}$. Outside the Lamb–Dicke regime these expressions break down. As n increases the motional wave-packet size becomes comparable to the wavelength of the driving field. This causes the amplitude of the driving field averaged over the wave-packet to decrease, giving a reduction in carrier Rabi frequency, as well as causing the gradient of the driving field over the wave-packet to decrease leading to a reduction in sideband Rabi frequency. This can be expressed analytically [WMI+98]

$$\left|\langle n'|\exp\left(i\eta(a + a^\dagger)\right)|n\rangle\right| = e^{-\eta^2/2} \sqrt{\frac{n_<!}{n_>!}} \eta^{|n'-n|} L_{n_<}^{|n'-n|}(\eta^2) \tag{2.13}$$

Fig. 2.4 Coupling strength to carrier and first motional sideband for $\eta = 0.15$. The *solid lines* show the full solution, and the *dashed lines* the solution in the Lamb–Dicke approximation

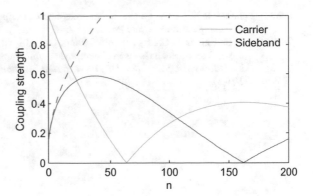

where $L_n^\alpha(x)$ is the generalised Laguerre polynomial, and $n_<$ ($n_>$) is the smaller (larger) of n and n'. This is plotted for a mode with $\eta = 0.15$ in Fig. 2.4. We see that the $\Delta n = 1$ matrix element deviates from the Lamb–Dicke regime expression $\sqrt{n}$, dropping to zero at $n \sim 160$.

References

[LRH+04]   Lucas, D.M., A. Ramos, J.P. Home, M.J. McDonnell, S. Nakayama, J.-P. Stacey, S.C. Webster, D.N. Stacey, and A.M. Steane. 2004. Isotope-selective photoionization for calcium ion trapping. *Physical Review A* 69 (1): 1–13.

[GRB+01]   Gulde, S., D. Rotter, P. Barton, F. Schmidt-Kaler, R. Blatt, and W. Hogervorst. 2001. Simple and efficient photo-ionization loading of ions for precision ion-trapping experiments. *Applied Physics B* 73 (8): 861–863.

[Szw09]   Szwer, D.J. 2009. *High Fidelity Readout and Protection of a 43Ca+ Trapped Ion Qubit.* PhD thesis, University of Oxford.

[Har13]   Harty, T.P. 2013. *High-Fidelity Microwave-Driven Quantum Logic in Intermediate-Field 43Ca+.* PhD thesis.

[Jan14]   Janacek, H.A. 2014. *Simulating trapped-ion qubits using the optical Bloch equations.* PhD thesis, University of Oxford.

[Deh75]   Dehmelt, H. 1975. Proposed $10^{14}\delta\nu > \nu$ laser fluorescence spectroscopy of Tl$^+$ mono-ion oscillator II. *Bulletin of the American Physical Society* 20 (1): 60.

[MSW+08]   Myerson, A., D.J. Szwer, S.C. Webster, D.T.C. Allcock, M.J. Curtis, G. Imreh, J.A. Sherman, D.N. Stacey, A.M. Steane, and D.M. Lucas. 2008. High-fidelity readout of trapped-ion qubits. *Physical Review Letters* 100 (20): 200502.

[Bur10]   Burrell, A.H. 2010. *High-fidelity readout of trapped-ion qubits.* PhD thesis, University of Oxford.

[MSW+04]   McDonnell, M.J., J.-P. Stacey, S.C. Webster, J.P. Home, A. Ramos, D.M. Lucas, D.N. Stacey, and A.M. Steane. 2004. High-efficiency detection of a single quantum of angular momentum by suppression of optical pumping. *Physical Review Letters* 93 (15): 1–4.

[LL76]   Landau, L.D. and E.M. Lifshitz. 1976. *Mechanics (3rd Ed.)*, pages 93–95. Butterworth-Heinemann.

[BMB+98]   Berkeland, D.J., J.D. Miller, J.C. Bergquist, W.M. Itano, and D.J. Wineland. Minimization of ion micromotion in a Paul trap. *Journal of Applied Physics*, 83(10), 1998.

[Jam98]   James, D.F.V. 1998. Quantum dynamics of cold trapped ions with application to quan-
             tum computation. *Applied Physics B: Lasers and Optics* 66 (2): 181–190.
[WMI+98]   Wineland, D.J., C. Monroe, W.M. Itano, D. Leibfried, B.E. King, and D.M. Meekhof.
             1998. Experimental issues in coherent quantum-state manipulation of trapped atomic
             ions. *Journal Of Research Of The National Institute Of Standards And Technology*,
             103(3).

Chapter 3
Raman Interactions

We use two-photon Raman interactions for the majority of our coherent qubit interactions. In this chapter we begin by deriving the Raman coupling in a model system, and then describe the undesired photon scattering (incoherent couplings) that produce a considerable portion of the error in our experiments. We then calculate the coupling rates for calcium, and conclude with experimental verification of these models.

3.1 The Raman Coupling

We start by considering a three state system consisting of levels $|1\rangle, |2\rangle$, and $|e\rangle$ (see Fig. 3.1) interacting with a pair of travelling wave fields. We drive the transitions $|1\rangle \leftrightarrow |e\rangle$ and $|2\rangle \leftrightarrow |e\rangle$ with Rabi frequencies, detunings, phases, and wave-vectors $\Omega_1, \delta_1, \phi_1, \mathbf{k}_1$ and $\Omega_2, \delta_2, \phi_2, \mathbf{k}_2$. The Hamiltonian for this system, in the interaction picture, is

$$\mathcal{H} = \frac{1}{2}\Omega_1 e^{i\phi_1} e^{i\mathbf{k}_1 \cdot \mathbf{r} - i\delta_1 t} |e\rangle \langle 1| + \frac{1}{2}\Omega_2 e^{i\phi_2} e^{i\mathbf{k}_2 \cdot \mathbf{r} - i\delta_2 t} |e\rangle \langle 2| + \text{h.c.} \quad (3.1)$$

If $\Omega_1 \ll |\delta_1|$ and $\Omega_2 \ll |\delta_2|$ we would normally consider all the terms in this Hamiltonian to be sufficiently far off resonant to be impotent. What we show in the following is that if $\delta_1 = \delta_2$ a two-photon effect causes a coupling between $|1\rangle$ and $|2\rangle$ – this is the Raman interaction we are interested in.

Assuming $|\delta_2 - \delta_1| \ll \delta_1, \delta_2$ we can apply the James–Jerk approximation [JJ07]. This gives us an effective Hamiltonian of

© Springer International Publishing AG 2017
C.J. Ballance, *High-Fidelity Quantum Logic in Ca+*,
Springer Theses, DOI 10.1007/978-3-319-68216-7_3

Fig. 3.1 Model system
demonstrating a Raman
transition. We couple the
states $|1\rangle$ and $|2\rangle$ by driving
the transitions $|1\rangle \leftrightarrow |e\rangle$ and
$|2\rangle \leftrightarrow |e\rangle$

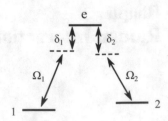

$$\mathcal{H} = \frac{\Omega_1^2}{4\delta_1}\left(|1\rangle\langle 1| - |e\rangle\langle e|\right) + \frac{\Omega_2^2}{4\delta_2}\left(|2\rangle\langle 2| - |e\rangle\langle e|\right)$$

$$+\frac{\Omega_1\Omega_2}{4\Delta}\left(e^{i\phi}e^{i\Delta k - i\delta t}|2\rangle\langle 1| + \text{h.c.}\right) \tag{3.2}$$

where $\phi := \phi_1 - \phi_2$, $\delta := \delta_1 - \delta_2$, $\Delta k := k_1 - k_2$, and $1/\Delta := \frac{1}{2}(\frac{1}{\delta_1} + \frac{1}{\delta_2})$. The first
two terms in this Hamiltonian are light-shifts. The final term is the Raman coupling –
this term coherently couples $|1\rangle$ and $|2\rangle$ with a coupling detuning and phase given by
the difference in the detunings and phases of the two beams. Ignoring the light-shifts
for the time being, the form of this effective coupling is exactly that of our standard
pseudo-spin coupling (Eq. 2.2), with $k \to \Delta k$, $\phi \to \phi_1 - \phi_2$, and $\delta \to \delta_1 - \delta_2$.

In a typical experimental implementation $|1\rangle, |2\rangle$ are states of the same parity
in the ground level S and $|e\rangle$ is an opposite parity state in the P level. Thus the
Raman coupling allows us to drive a (low frequency) transition between ground state
levels of the same parity with optical photons. The advantages over driving the low
frequency transition directly (e.g. with RF or microwaves) are the large momentum
of optical photons versus RF photons (allowing strong spin-motion coupling) and
the ability to tightly focus optical beams ($\lambda \sim 400$ nm) versus RF/microwave beams
($\lambda \sim 10$ m–10 cm), allowing individual ion addressing.

3.2 Rabi Frequencies and Light-Shifts

In a typical system we couple the two states in the ground level via many excited
states $\{|e_i\rangle\}$. As long as the 'far detuned' approximation of the previous section holds
for each of the excited states we can (coherently) sum the Rabi frequencies from each
path. Thus the net effective Rabi frequency is

$$\Omega_{12} = \sum_i \frac{\langle 2| E_2\hat{\epsilon}_2 \cdot \boldsymbol{d} |e_i\rangle \langle e_i| E_1\hat{\epsilon}_1 \cdot \boldsymbol{d} |1\rangle}{2\Delta_i} \tag{3.3}$$

where $E_{1,2}$ are the electric field magnitudes at the atom, $\hat{\epsilon}_{1,2}$ are the beam polarisations, $\boldsymbol{d}$ is the dipole operator, and Δ_i is the mean detuning of the beams from
excited state $|e_i\rangle$ (as previously defined). We define the 'coupling strength' of a beam

as $g := \mu E$, where μ is the largest matrix element connecting the ground level to the excited level (see Sect. A.2) – when multiplied by an angular factor this gives the on-resonance Rabi frequency. Rewriting our Rabi frequency

$$\Omega_{12} = \frac{g_1 g_2}{2} \sum_i \frac{\langle 2| \, \hat{\epsilon}_2 \cdot \boldsymbol{d} \, |e_i\rangle \, \langle e_i| \, \hat{\epsilon}_1 \cdot \boldsymbol{d} \, |1\rangle}{\mu^2 \Delta_i} \tag{3.4}$$

The light-shifts on a ground level state, for example $|1\rangle$, from a single beam in the same notation are

$$\mathcal{H} = \left(\sum_i \frac{\langle e_i| \, \hat{\epsilon} \cdot \boldsymbol{d} \, |1\rangle^2}{4\mu^2 \Delta_i} \right) |1\rangle \langle 1| \tag{3.5}$$

3.3   Photon Scattering

The two processes we have looked at so far, Raman transitions and light-shifts, are coherent processes caused by stimulated absorption from and emission into the applied laser fields. However the excited states $\{|e_i\rangle\}$ we couple through decay spontaneously, hence there are also processes with vacuum fluctuation driven emission in place of the second field term. The two possible processes are Raman scattering, an inelastic process where the final atomic state is different from the initial atomic state, and Rayleigh scattering, where the final state and initial state are the same.

The scattering rate from $|i\rangle$ to $|f\rangle$ is calculated from the Kramers–Heisenburg formula to be [CMMH94, OLJ+05]

$$\Gamma_{i,f} = \frac{g^2 \gamma}{4} \sum_k \left| \sum_{e,q} \frac{\langle f| \hat{\epsilon}_q \cdot \boldsymbol{r} \, |e\rangle \, \langle e| \, b_k \hat{\epsilon}_k \cdot \boldsymbol{r} \, |i\rangle}{\mu^2 \Delta_e} \right|^2 \tag{3.6}$$

where q is the emitted photon's polarisation, b_k are the applied field polarisation components, g is the applied field coupling strength, and γ is the excited state A coefficient.

The Raman and Rayleigh parts of the scattering rate Γ_i out of state $|i\rangle$ are $\Gamma_{\text{Rayleigh}} := \Gamma_{i,i}$ and $\Gamma_{\text{Raman}} := \sum_{f \neq i} \Gamma_{i,f}$, where $\Gamma_i = \Gamma_{\text{Rayleigh}} + \Gamma_{\text{Raman}}$.

3.3.1   Raman Scattering

The Raman scattering terms change the atom's internal state, hence the scattered photons are entangled with the atom's internal state. The environment 'measures' these photons leading to decoherence of the internal state. The atom's motion is also decohered by the recoil from the photon emission in a random direction.

3.3.2   Rayleigh Scattering

The Rayleigh scattering process is elastic. If the Rayleigh scattering amplitudes for the two qubit states are the same the atom's internal state is separable from the scattered photon's state. In this case the Rayleigh scattering events do not decohere the *internal* state of the atom. The Rayleigh scattering events still transfer momentum to the atom and thus decohere the motion.

If the Rayleigh scattering amplitudes for the two qubit states differ then Rayleigh scattering dephases the qubit [UBV+10]. The dephasing rate is given by

$$\Gamma_{el} = \frac{g^2\gamma}{4} \sum_k \left| \sum_{e,q} \frac{\langle \uparrow | \hat{\epsilon}_q \cdot \boldsymbol{r} | e \rangle \langle e | b_k \hat{\epsilon}_k \cdot \boldsymbol{r} | \uparrow \rangle}{\mu^2 \Delta_e} - \frac{\langle \downarrow | \hat{\epsilon}_q \cdot \boldsymbol{r} | e \rangle \langle e | b_k \hat{\epsilon}_k \cdot \boldsymbol{r} | \downarrow \rangle}{\mu^2 \Delta_e} \right|^2 \tag{3.7}$$

where $|\downarrow\rangle$ and $|\uparrow\rangle$ are the two qubit states. This dephasing gives rise to Lindblad terms in the master equation like

$$L_{el} = \sqrt{\frac{\Gamma_{el}}{4}} \sigma_z \tag{3.8}$$

3.3.3   Motional Decoherence from Scattering

In both the Raman and Rayleigh scattering processes the atom absorbs a photon from the Raman beam and emits a photon in a random direction. The total momentum transfer produces a displacement of $D(\Delta k z_0)$ where z_0 is the ground state wave packet size, Δk is the difference in k-vector between the absorbed and emitted photon, and $D(\alpha)$ is the motional displacement operator. Starting in the motional ground state $|n = 0\rangle$ this excites us to $|n = 1\rangle$ with probability $|\Delta k z_0|^2 \sim \eta^2 \ll 1$. For Raman scattering this small probability of motional error is dwarfed by the certain error from the change in internal state, so we ignore it. For Rayleigh scattering this could potentially be the leading source of error. As a very crude model we can assume this gives us an effective motional heating rate of $\eta^2 \Gamma_{\text{Rayleigh}}$ – this is a very small effect that is completely negligible in the experiments in this thesis. The effect of this process on two-qubit gates is considered in more detail in Sect. III of [OIB+07] with similar results.

Fig. 3.2 Simplified level
scheme of ^{43}Ca$^+$. The
Raman beams address the
S-P transition. The P states
decay mostly back to S, at
rate $\gamma = 132 \times 10^6\mathrm{s}^{-1}$, and
occasionally into D at rate
$\alpha\gamma$ ($\alpha \approx 0.06$). The
hyperfine splittings are
shown for the $S_{1/2}$ and $P_{1/2}$
levels only

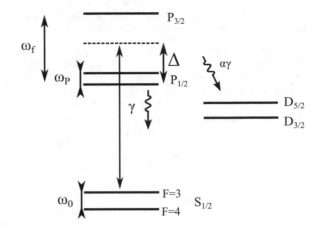

3.4   Raman Transitions in ^{43}Ca$^+$

A simplified level structure of ^{43}Ca$^+$ is shown in Fig. 3.2. The Raman beams address
the S-P transition ($\approx$395 nm), with the Raman detuning Δ defined as the detuning
of the applied beams from the $S_{1/2}^{4,+4} \leftrightarrow P_{1/2}^{4,+4}$ transition. The fine structure splitting
between $P_{1/2}$ and $P_{3/2}$ is $\omega_f = 2\pi \cdot 6.68$ THz. As this is comparable to our typical
Raman detunings we couple significantly to both P levels.

As the P states can decay to the D states we should also consider the Raman
scattering rate from S to D in our calculations. As the branching ratio to the D
states, α, is small we often ignore this – thus in this thesis we will generally use the
term 'Raman scattering' to refer only to scattering back to S. When we consider the
Raman scattering to D we will explicitly state this. The scattering rate to D is given
by $\alpha\Gamma_{\text{total}}$, where Γ_{total} is the total scattering rate on S $\leftrightarrow$ P.

In calculating the Raman rates we assume $|\Delta| \gg \omega_0, \omega_P$. For the single-beam
Raman rates, the beam coupling strength is given by g, and the beam polarisation
amplitudes are given by $\{e_-, e_0, e_+\}$ (where $e_-^2 + e_0^2 + e_+^2 = 1$). For the two-beam
Raman rates, the beam coupling strengths are given by g_r and g_b (for the lower
frequency 'red' beam and the higher frequency 'blue' beam respectively), with the
polarisations given by $\{r_-, r_0, r_+\}$ and $\{b_-, b_0, b_+\}$.

3.4.1   Stretch Qubit

The majority of the work in this thesis was performed on the magnetically sensitive
'stretch' qubit ($|\downarrow\rangle := S_{1/2}^{4,+4}$ and $|\uparrow\rangle := S_{1/2}^{3,+3}$). The Raman rates relevant to the
work in this thesis are given in Table 3.1. As we will see in Sect. 3.5.2 the Rayleigh
scattering induced dephasing is a significant proportion of the total photon scattering
error.

Table 3.1 Stretch qubit Raman rates. The light-shift, dephasing rate, and scattering rates are all calculated for a single beam with coupling strength g. The 'light-shift force' will be defined in Chap. 4

Rabi frequency	$\Omega_{\uparrow\downarrow} = \frac{\sqrt{7}}{12}(b_- r_0 + b_0 r_+) g_r g_b \left[\frac{\omega_f}{\Delta(\Delta - \omega_f)} \right]$
Differential light-shift	$\Delta_{\text{LS}} = \frac{7}{48}(e_+^2 - e_-^2) g^2 \left[\frac{\omega_f}{\Delta(\Delta - \omega_f)} \right]$
Elastic dephasing rate	$\Gamma_{\text{el}} = \frac{49}{576}(e_+^2 + e_-^2) \gamma g^2 \left[\frac{\omega_f}{\Delta(\Delta - \omega_f)} \right]^2$
Light-shift force ($\sigma_\pm \perp \sigma_\pm$)	$\Omega_\downarrow = -\frac{1}{6} g_r g_b \left[\frac{\omega_f}{\Delta(\Delta - \omega_f)} \right]$
	$\Omega_\uparrow = \frac{1}{8} g_r g_b \left[\frac{\omega_f}{\Delta(\Delta - \omega_f)} \right]$
Raman scattering rate with $\sigma_\pm$ light	$\Gamma_{\text{Raman}}^\downarrow = \frac{1}{36} \gamma g^2 \left[\frac{\omega_f}{\Delta(\Delta - \omega_f)} \right]^2$
	$\Gamma_{\text{Raman}}^\uparrow = \frac{23}{576} \gamma g^2 \left[\frac{\omega_f}{\Delta(\Delta - \omega_f)} \right]^2$
Total scattering rate	$\Gamma_{\text{Total}} = \frac{1}{12} \gamma g^2 \left[\frac{1}{\Delta^2} + \frac{2}{(\Delta - \omega_f)^2} \right]$

Table 3.2 Low-field clock qubit Raman rates. The light-shift and scattering rates are all calculated for a single beam

Rabi frequency	$\Omega_{\uparrow\downarrow} = \frac{1}{6}(b_- r_- - b_+ r_+) g_r g_b \left[\frac{\omega_f}{\Delta(\Delta - \omega_f)} \right]$
Differential light-shift	$\Delta_{\text{LS}} = \frac{1}{12} g^2 \left[\frac{\omega_0}{\Delta(\Delta + \omega_0)} + \frac{2\omega_0}{(\Delta - \omega_f)(\Delta - \omega_f + \omega_0)} \right]$
Elastic dephasing rate	$\Gamma_{\text{el}} = 0$
Raman scattering rate	$\Gamma_{\text{Raman}} = \frac{1}{18} \gamma g^2 \left[\frac{\omega_f}{\Delta(\Delta - \omega_f)} \right]^2$
Total scattering rate	$\Gamma_{\text{Total}} = \frac{1}{12} \gamma g^2 \left[\frac{1}{\Delta^2} + \frac{2}{(\Delta - \omega_f)^2} \right]$
Transitions out of clock state ($\sigma_\pm$)	$\Omega_{4,0-4,\pm2} = \Omega_{3,0-4,\pm2} =$ $\frac{\sqrt{5}}{32\sqrt{2}} g_b g_r \left[\frac{\omega_P}{\Delta(\Delta - \omega_P)} \right]$
	$\Omega_{4,0-3,\pm2} = \Omega_{3,0-3,\pm2} =$ $\frac{\sqrt{5}}{32\sqrt{6}} g_b g_r \left[\frac{\omega_P}{\Delta(\Delta - \omega_P)} \right]$

3.4.2   Low-Field Clock Qubit

The low-field clock qubit is the pair of states $|\downarrow\rangle := S_{1/2}^{4,0}$ and $|\uparrow\rangle := S_{1/2}^{3,0}$. The relevant Raman rates are given in Table 3.2. An interesting observation is that for beam polarisations that drive the clock qubit transition optimally (no π component) the Rabi frequencies of transitions leading out of the clock qubit are negligible: the ratio of the out-of-qubit transition rate to the qubit transition rate is $\sim 10^{-4}$ at

$\Delta = -1$ THz. This means we can drive this qubit very fast without worrying about off-resonant excitation – this is very useful for the work described in Chap. 7.

3.4.3 Scaling of Raman Rates with Detuning

The sources of error in driving Raman transitions, namely Raman scattering and the dephasing from Rayleigh scattering, decrease faster with an increase in detuning Δ than the Raman Rabi frequency. This means that the probability of error, the ratio of error rate to Rabi frequency, decreases for increased detuning. We can thus choose between operation speed and operation error for a given beam intensity by adjusting the detuning.

Figure 3.3 is a plot of $\Gamma_{\text{Raman}}/\Omega_{\uparrow\downarrow}$ and $\Gamma_{\text{Tot}}/\Omega_{\uparrow\downarrow}$, the ratio of Raman scattering rate and total scattering rate to Rabi frequency. This particular plot is for the stretch qubit, however the qualitative structure is the same for the clock qubit (as can be seen from the scaling with detuning of the rates in Tables 3.1 and 3.2). The Raman scattering to Rabi frequency ratio asymptotically approaches zero for large detunings, thus the error from Raman scattering can always be reduced by detuning further. The Rayleigh scattering to Rabi frequency ratio, however, asymptotically approaches $\sim \gamma/\omega_f$. This means that in the limit of large detuning a constant number of Rayleigh photons are scattered per operation. The rate of scattering into the D states is proportional to the total number of scattered photons on the S-P transition, so also has an asymptotic limit. This limits the ultimately attainable operation fidelity in ions with low-lying D states [OIB+07]. In the case of ^{43}Ca$^+$ this limit is $\approx 2 \times 10^{-6}$ for a carrier π-pulse.

We can qualitatively understand these asymptotic limits by considering the energy scales of the atomic structure. The laser beams we apply only couple to the electron position. To use a laser beam to flip the electron or nuclear spin we reply on the spin-orbit or hyperfine coupling respectively. If we are far detuned compared to the splitting (i.e. $|\Delta| \gg \omega_{\text{split}}$) the coupling 'averages out'. For detunings $|\Delta| \gg$

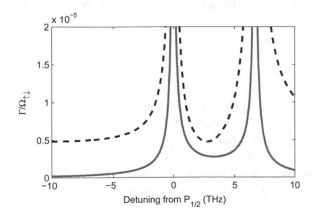

Fig. 3.3 Ratio of the scattering rates to the Raman Rabi frequency, evaluated for the stretch states. The *solid blue line* is calculated using the rate of Raman scattering alone, and the *dashed black line* using the total scattering rate (Raman+Rayleigh) (color figure online)

$\omega_{HF} \sim$GHz the hyperfine coupling has little effect on the Raman transition rates – only the electron spin can be flipped. This is similar to transitions directly driven by microwaves. Thus for Raman transitions between states in the ground level the microwave transition matrix elements are very similar to the Raman transition matrix elements (appendix A.3). This means that one cannot use the two-photon Raman process to drive $\delta m_f = 2$ transitions – these transitions are suppressed by ω_{HF}/Δ, for typical detunings of $\sim$THz this suppression is 10^{-3}. For detunings $|\Delta| \gg \omega_f$ the spin-orbit coupling becomes negligible, leading to a decoupling of the electron spin and position, and hence the asymptotic limit of the ratio of photon scattering to Rabi frequency.

In any experiment we are technically limited in the maximum laser beam intensity we can apply. For a given operation error we have a choice of detunings; in the far detuned case we can choose to tune red or blue of the P levels, for more moderate detunings we can also choose to detune in between the P levels, either closer to $P_{1/2}$ or $P_{3/2}$. It is natural to ask if any of these choices give a higher Rabi frequency for a given intensity. Considering only Raman scattering back to the S level the different detuning choices are all identical. Including Raman scattering to the D levels an asymmetry emerges; when possible it is most efficient to tune inside the P manifold towards $P_{1/2}$, for lower error rates no solution exists inside the manifold and the most efficient solution is to tune below $P_{1/2}$. The difference in efficiency is relatively small, about 10% in the beam powers required.

3.5 Scattering Experiments

To confirm that our Rabi frequency and photon scattering models are correct we perform several test experiments on the 'stretch' qubit of ^{43}Ca^{+}. In Sect. 3.5.1 we measure the ratio of the photon scattering rate to the qubit's differential light-shift, and in Sect. 3.5.2 we confirm our model of the Rayleigh dephasing of the qubit.

3.5.1 Ratio of Scattering Rate to Light-Shift

To compare our scattering rate calculations with our experimental results we need to know the intensity of the scattering beam at the ion. This is difficult to determine accurately by direct measurement of the beam power and spot size. However the ratio of the scattering rate to the differential light-shift of the qubit does not depend on the beam intensity. Assuming we know the polarisation and the detuning of the applied beam, and that the applied beam is monochromatic, we have a parameter-free comparison of the experimental results with the theory.

In our experiment we apply a σ^+ polarised beam detuned below the $S_{1/2} \leftrightarrow P_{1/2}$ transition for a varying period of time to the initial state $S_{1/2}^{3,+3}$ and measure the population in $F = 4$. The population is slowly pumped to $S_{1/2}^{4,+4}$ with only one level outside the qubit ($S_{1/2}^{4,+3}$) being involved. Due to the (small) probability of scattering out of $S_{1/2}^{4,+3}$ to $S_{1/2}^{3,+3}$ the population of the $F = 4$ manifold is described by a bi-exponential. However numerical simulations show that to $\approx 1\%$ accuracy the $F = 4$ population can be described by

$$P_{F=4} = 1 - e^{-\Gamma t}$$
$$\Gamma := \Gamma_{3,+3\to4,+3} + \Gamma_{3,+3\to4,+4} \qquad (3.9)$$

assuming $|\Delta| \gg \omega_0$; our theory thus predicts

$$\frac{\Gamma}{\Delta_{LS}} = \frac{5}{12}\gamma \left| \frac{\omega_f}{\Delta(\Delta - \omega_f)} \right| \qquad (3.10)$$

The beam we use for this experiment has a $\sim 10\%$ σ^- impurity. We can extract this beam polarisation accurately by measuring the equilibrium state after applying the scattering beam for a long time. With the beam polarisation known we can use our numerical model to calculate what the measured ratio of scattering rate to light-shift should be. From the equilibrium spin state we infer a polarisation of $e = \{\sigma^-, \pi, \sigma^+\} = \{0.2, 0, 0.98\}$.

Figure 3.4 shows the experimentally measured scattering rate to light-shift ratio alongside the theoretical prediction (including the polarisation correction). The agreement is good for $|\Delta| > 100$ GHz, but for detunings smaller than this the measured ratio is somewhat larger than predicted – this divergence is due to the neglected hyperfine splittings.

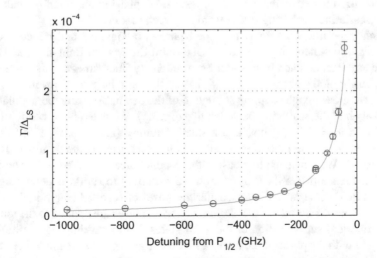

Fig. 3.4 Measured and theoretical ratio of Raman scattering rate to differential light-shift for nominally σ^+ light on the stretch qubit. The detuning errors are much smaller than the symbol size

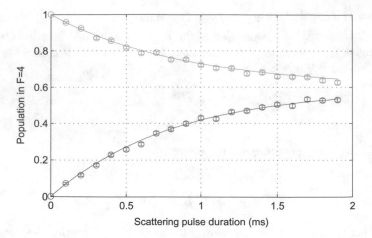

Fig. 3.5 Population after a $\sigma_\pm$ scattering pulse, starting from each of the stretch qubit states. The *solid lines* are the predictions of the Raman scattering model (with the beam coupling strength, g, the only free parameter)

3.5.2 Stretch Qubit Rayleigh Dephasing

To confirm our calculations of the dephasing caused by Rayleigh scattering (Eq. 3.8, derived in [UBV+10]) we perform an additional experiment. We measure the rate of scattering out of each of the stretch qubit states and calculate from this the expected rate of decoherence of the superposition $|\downarrow\rangle + |\uparrow\rangle$ from Raman scattering alone. We then measure the decoherence rate; the 'excess' dephasing on top of the calculated Raman scattering dephasing rate is from Rayleigh scattering.

In this experiment we apply a single beam with equal σ_+ and σ_- components (and no π component) in order to null the qubit differential light-shift, and hence remove any decoherence from small beam intensity fluctuations. This complicates the scattering model as population is scattered through the ground level.

First we measure the coupling strength g of the beam, by fitting the population in $F = 4$ against the scattering pulse length (Fig. 3.5). We use a large beam intensity (5 mW in a $w = 27\,\mu$m spot) and a small detuning ($\Delta = -2\pi \cdot 138.2$ GHz) to give a large scattering rate, to ensure the photon scattering is the dominant form of decoherence. We now need to measure the decoherence rate. We use a fixed length CPMG sequence [MG58] (using microwave π-pulses) to suppress dephasing from magnetic field noise, with photon scattering pulses of scanned length in the CPMG sequence delays. This means the photon scattering pulses are applied to the ion when it is in an equal superposition of the qubit states. By measuring the resulting spin population with the phase of the final CPMG pulse at 0 and π we sample the Ramsey fringe contrast, which yields the density matrix element $\rho_{\uparrow,\downarrow}$ and hence measures the coherence remaining after scattering. The measured coherence and the results of the model excluding and including Rayleigh dephasing are shown in Fig. 3.6. The

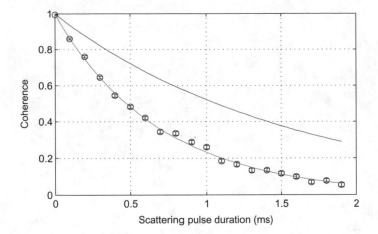

Fig. 3.6 Decoherence of the stretch qubit after a $\sigma_\pm$ polarised scattering pulse. The *blue line* is the model result if the only source of decoherence is Raman photon scattering. The *green line* is the model result including dephasing from Rayleigh scattering

measured dephasing rate is roughly a factor of two higher than would result from Raman scattering alone, showing that Rayleigh dephasing is a significant effect. The model including the Rayleigh dephasing is in very good agreement with the experiment.

References

[JJ07] James, D.F.V., and J. Jerke. 2007. Effective Hamiltonian theory and its applications in quantum information. *Canadian Journal of Physics* 85 (6): 625–632.

[CMMH94] Cline, R.A., J.D. Miller, M.R. Matthews, and D.J. Heinzen. 1994. Spin relaxation of optically trapped atoms by light scattering. *Optics letters* 19 (3): 207.

[OLJ+05] Ozeri, R., C.E. Langer, J.D. Jost, B. DeMarco, A. Ben-Kish, R.B. Blakestad, J.W. Britton, J. Chiaverini, W.M. Itano, D.B. Hume, D. Leibfried, T. Rosenband, P.O. Schmidt, and D.J. Wineland. 2005. Hyperfine Coherence in the Presence of Spontaneous Photon Scattering. *Physical Review Letters* 95 (3): 1–4.

[UBV+10] Uys, H., M.J. Biercuk, A.P. VanDevender, C. Ospelkaus, D. Meiser, R. Ozeri, and J.J. Bollinger. 2010. Decoherence due to Elastic Rayleigh Scattering. *Physical Review Letters* 105 (20): 200401.

[OIB+07] Ozeri, R., W.M. Itano, R.B. Blakestad, J.W. Britton, J. Chiaverini, J.D. Jost, C.E. Langer, D. Leibfried, R. Reichle, S. Seidelin, J.H. Wesenberg, and D.J. Wineland. 2007. Errors in trapped-ion quantum gates due to spontaneous photon scattering. *Physical Review A* 75 (4): 1–14.

[MG58] Meiboom, S., and D. Gill. 1958. Modified Spin-Echo Method for Measuring Nuclear Relaxation Times. *Review of Scientific Instruments* 29 (8): 688.

Chapter 4
Two-Qubit Gate Theory

In this chapter we give an overview of the implementation of two-qubit gates in trapped-ion systems. We explain the concept behind the 'geometric phase' gate that is commonly used. We then discuss our specific implementation of this gate mechanism, the 'light-shift' gate. The chapter concludes with an analysis of the different sources of error in the light-shift two-qubit gate mechanism.

4.1 Introduction

If one wants to implement universal quantum computation one needs to be able to implement unitary operations spanning a large Hilbert space (many qubits). It can be shown that from a set of single-qubit operations and a single two-qubit operation that one can build up these desired arbitrary operations. As multi-qubit operations are far more difficult than single operations this represents a great simplification - we 'just' need to implement one two-qubit operation with high fidelity alongside our set of single-qubit operations. Although in this chapter we will only consider gates on two qubits, the methods described all work on many qubits with little extra (theoretical) complexity.

We wish to generate a coupling between the internal states of the two ions. The ions' internal (electronic) states do interact directly via the magnetic spin-spin interaction, but as the typical inter-ion spacing in an ion crystal is $5\,\mu m$ this direct interaction is very weak (mHz; nevertheless it has recently been measured [KAN+14]). The Coulomb interaction between the ions, however, is strong ($\sim$MHz); as we have seen in Sect. 2.3 this leads to shared modes of motion of the ion crystal. In the following we will discuss techniques to use this motional coupling as a 'bus' through which the two ions communicate their spin state.

© Springer International Publishing AG 2017
C.J. Ballance, *High-Fidelity Quantum Logic in Ca+*,
Springer Theses, DOI 10.1007/978-3-319-68216-7_4

There have been many schemes proposed to implement entangling gates mediated by the motion in ion traps [CZ95, SrM199, JPK00, vS03, LDM+03, GRZC03, BSPR12]. The most experimentally successful implementations all share a common mechanism – they operate by generating a spin-dependent geometric phase.

4.2 The Geometric Phase Gate

Geometric phase gates rely on the use of a spin-dependent force to drive the spin states around closed paths in phase space. When these paths close the spin and motion are separable, but path-dependent phase shifts occur – this is the Berry geometric phase [Ber84].[1] This is the mechanism behind the operation of the light-shift gate [LDM+03], Mølmer-Sørensen gate [SrM199], and Bermudez gate [BSPR12], amongst others, the only difference being the basis of and the mechanism used to generate the spin-dependent force.

We start by calculating the (interaction picture) Hamiltonian for a forced harmonic oscillator [Car65] with forcing frequency ω

$$
\begin{aligned}
\mathcal{H}_{\text{force}} &= -zF(t) \\
&= -z_0 F_0 (ae^{-i\omega_z t} + a^\dagger e^{+i\omega_z t}) \cos(\omega t - \phi) \\
&= \frac{z_0 F_0}{2} a^\dagger e^{i\phi - i\delta t} + \text{h.c.}
\end{aligned}
\tag{4.1}
$$

where ω_z is the harmonic oscillator frequency, and we have assumed $\delta := \omega - \omega_z \ll \omega_z$ and dropped far off-resonant terms. As we will see there are several ways to experimentally generate such a term with a force amplitude F_0 conditioned on the spin state.

We now assume that we have a set of spins coupled to a single motional mode by a spin-dependent force

$$
\mathcal{H}_{\text{SDF}} = \frac{i\Omega_D}{2} \Lambda a^\dagger e^{i\phi} e^{-i\delta t} + \text{h.c.}
\tag{4.2}
$$

where Λ is some combination of Pauli matrices acting on all the spins, and Ω_D is the coupling Rabi frequency. Using a Magnus expansion [Mag54] we can exactly integrate this Hamiltonian to get the propagator

---

[1] We note that the total accumulated phase has a dynamic as well as a geometric component, however the total phase is proportional to the geometric phase, hence the total phase is proportional to the area encircled in phase space [Oze11].

$$U(t) = D(\alpha(t)\Lambda)e^{-i\Phi(t)\Lambda^2}$$

$$\alpha(t) = \frac{\Omega_D}{\delta} \sin\left(\frac{\delta t}{2}\right) e^{i\phi} e^{-i\delta t/2}$$

$$\Phi(t) = \frac{\Omega_D^2}{4\delta^2}(\delta t - \sin \delta t) \tag{4.3}$$

where $D(\cdot)$ is the displacement operator, defined $D(\alpha) := \exp(\alpha a^\dagger - \alpha^* a)$. We can make the operation of this propagator more obvious by expanding in the eigenbasis of the operator Λ

$$U = \sum_\lambda D(\alpha(t)\lambda)e^{-i\Phi(t)\lambda^2} |\lambda\rangle \langle\lambda| \tag{4.4}$$

where $\{\lambda\}$ are the eigenvalues of Λ. We can see that we displace the eigenstates of Λ dependent on the phase and magnitude of the eigenvalue, and that each eigenstate gains a geometric phase $\Phi(t)\lambda^2$. If we apply this propagator for an arbitrary time we will, in general, have entangled the spin and motion. When we then measure the spin only (and hence trace out the motion) we lose coherence. If, however, we choose $\tau = 2\pi K/\delta$ (with integer K), we find $\alpha(\tau) = 0$, and hence

$$U(\tau) = \sum_\lambda \exp\left(-i\Phi(\tau)\lambda^2\right) |\lambda\rangle \langle\lambda| \tag{4.5}$$

In this case we have generated a phase which is dependent on the system spin state, while leaving the spin-state separable from the motion. Our spin-dependent force has driven different parts of the spin state around different (circular) paths in the motional phase space (described by the displacement α). In doing this we accumulate a geometric phase (Φ) that is dependent on the area enclosed by the motional path. When $\alpha = 0$ (i.e. $\delta\tau = 2\pi K$) we have completed K loops in phase space. This propagator is independent of the initial state of the motion. We will see later on that the interaction Hamiltonian we can generate is not exactly that of Eq. 4.2 and that this causes a dependence on the initial motional state.

As an explicit example of how the propagator of Eq. 4.3 can be used to implement a two-qubit gate we consider the interaction operator $\Lambda = \frac{1}{2}(\sigma_{z,1} + \sigma_{z,2})$ (this is essentially the interaction we discuss in Sect. 4.3). We first note that $\Lambda^2 = \frac{1}{2}(I + \sigma_{z,1}\sigma_{z,2})$. We can now expand the propagator at times when $\alpha(\tau) = 0$

$$U(\tau) = [|\uparrow\downarrow\rangle \langle\uparrow\downarrow| + |\downarrow\uparrow\rangle \langle\downarrow\uparrow|] + \exp\left(-i\Phi(\tau)\right)[|\uparrow\uparrow\rangle \langle\uparrow\uparrow| + |\downarrow\downarrow\rangle \langle\downarrow\downarrow|] \tag{4.6}$$

If we choose $\Phi(\tau) = \frac{\pi}{2}$ we have implemented a symmetrized phase gate (up to a global phase)

$$U(\tau) = -i U_{\text{gate}} = -i \begin{pmatrix} 1 & & & \\ & i & & \\ & & i & \\ & & & 1 \end{pmatrix} \tag{4.7}$$

This can be mapped into the conventional controlled-not gate with single-qubit operations

$$U_{\text{CNOT}} = \begin{pmatrix} 1 & 0 & & \\ 0 & 1 & & \\ & & 0 & 1 \\ & & 1 & 0 \end{pmatrix} = (I \otimes H)(P \otimes P) U_{\text{gate}} (I \otimes H) \tag{4.8}$$

where H is the Hadamard operator and $P = \text{diag}(1, \ e^{-i\pi/2})$ is a single-qubit phase gate.

It will prove useful to have expressions for the spin state at all times, rather than just for times where $\alpha = 0$. We wrap the gate propagator (Eq. 4.3) in Hadamards, and apply it to $|\downarrow\downarrow\rangle$: for appropriate parameters this will generate the Bell state $|\downarrow\downarrow\rangle + |\uparrow\uparrow\rangle$. The spin populations, after tracing out the motion, are

$$P_{\uparrow\uparrow} = \frac{1}{8} \left(3 + e^{-4|\alpha|^2 (\bar{n} + \frac{1}{2})} - 4 \cos \Phi \, e^{-|\alpha|^2 (\bar{n} + \frac{1}{2})} \right)$$

$$P_{\uparrow\downarrow} + P_{\downarrow\uparrow} = \frac{1}{4} \left(1 - e^{-4|\alpha|^2 (\bar{n} + \frac{1}{2})} \right) \tag{4.9}$$

where we have assumed that the motion was initially in a thermal state with mean occupation number $\bar{n}$. The fidelity of the state produced with respect to the Bell state is

$$\mathcal{F} = \frac{1}{8} \left(3 + e^{-4|\alpha|^2 (\bar{n} + \frac{1}{2})} + 4 \sin \Phi \, e^{-|\alpha|^2 (\bar{n} + \frac{1}{2})} \right) \tag{4.10}$$

4.3 The Light-Shift Gate

In our experiments we implement the light-shift ('wobble') gate [LDM+03, HML+06]. In this scheme a pair of Raman beams with a frequency difference close to a motional mode frequency produce the force. The polarisation of the beams is chosen to couple unequally to the spin states, thus the magnitude and phase of the force are in general dependent on the spin state of the system.

The simplest way of understanding this force involves first considering the effect of only one Raman beam. If this beam couples to $|\downarrow\rangle$ differently from $|\uparrow\rangle$ there is a differential light-shift on the qubit. If we we now consider two Raman beams at the same frequency we get a stationary interference pattern. This means that the magnitude of the light-shift spatially varies, which gives, of course, a force. If we tune the beat frequency of the beams close to a motional mode frequency and neglect the off-resonant terms we have generated a spin-dependent force.

For this scheme to work we must be able to produce a differential light-shift between the qubit states. One can show that for 'clock' states the differential light-shift tends to zero for large Raman detunings ($|\Delta| \gg \omega_{HF}$) [LBD+05]. As we have seen in Chap. 3 we need a large Raman detuning to suppress photon scattering. This means that, in practice, one cannot use this light-shift gate on 'clock' qubits – a magnetically sensitive qubit transition is required.

Let us now consider the full Hamiltonian describing the light-shift gate. For simplicity we assume we have two ions, that the Raman beams are aligned to couple only to the axial modes, and that the Raman beams uniformly illuminate the ions. The Hamiltonian in the interaction picture with respect to the spin is

$$\mathcal{H} = \sum_j \frac{1}{2} \left(\Omega_\downarrow |\downarrow_j\rangle\langle\downarrow_j| + \Omega_\uparrow e^{i\delta\phi} |\uparrow_j\rangle\langle\uparrow_j| \right) e^{-i\omega t} e^{i\phi_0}$$

$$e^{i\Delta k z_j^0} e^{i\eta_c(a_c + a_c^\dagger)} e^{i\eta_b(-1)^j(a_b + a_b^\dagger)} + \text{h.c.} \qquad (4.11)$$

where $\delta\phi$ is the phase-shift between the force on $|\uparrow\rangle$ and $|\downarrow\rangle$, ω is the difference frequency between the two beams, j indexes the ions, ϕ_0 is the initial difference phase of the Raman beams whose difference wave-vector is Δk, and z_j^0 are the equilibrium positions of the ions. We have explicitly written the coupling to the centre-of-mass (c) and breathing (b) motional modes, and allowed for different force magnitudes ($\Omega_\uparrow$ and $\Omega_\downarrow$) on the different spin states.

First, let us see how this gives us a Hamiltonian similar to Eq. 4.2. We expand Eq. 4.11 in the Lamb–Dicke approximation, assuming that the beam polarisations are chosen to give $\delta\phi = \pi$, and that we are tuned near to resonance with the centre-of-mass mode ($\delta := \omega - \omega_z$, $|\delta| \ll \omega_z$), so we ignore excitation of the breathing mode

$$\mathcal{H} = \sum_j \frac{i\eta_c}{2} \left(\Omega_\uparrow |\uparrow_j\rangle\langle\uparrow_j| - \Omega_\downarrow |\downarrow_j\rangle\langle\downarrow_j| \right) e^{-i\delta t} e^{i\phi_0} e^{i\Delta k z_j^0} a_c^\dagger + \text{h.c.} \qquad (4.12)$$

This is a spin dependent force in the σ_z basis. For each of the spin states the Rabi frequencies that give the force phase and amplitude are:

$$\uparrow\uparrow : \ (1 + e^{i\phi_m})\Omega_\uparrow$$
$$\uparrow\downarrow : \ \Omega_\uparrow - e^{i\phi_m}\Omega_\downarrow$$
$$\downarrow\uparrow : \ e^{i\phi_m}\Omega_\uparrow - \Omega_\downarrow$$
$$\downarrow\downarrow : \ -(1 + e^{i\phi_m})\Omega_\downarrow \qquad (4.13)$$

where $\phi_m = \Delta k \cdot d$ is the Raman phase difference between the ions (d is the ion separation vector). As long as $\phi_m \neq \pi/2$ this spin-dependent force can be used to make a two-qubit entangling gate, however to maximise the efficiency of the force the Raman phase difference should be set to $\phi_m = \pi$: this produces no force on $\uparrow\uparrow$ or $\downarrow\downarrow$, and a force of amplitude $\pm(\Omega_\uparrow + \Omega_\downarrow)$ on $\uparrow\downarrow$ and $\downarrow\uparrow$. We discuss in Chap. 8 how we set this phase difference experimentally.

The phase difference between the force on spin state $\uparrow$ and on spin state $\downarrow$, $\delta\phi$, is set by the Raman beam geometry. Using a pair of beams in the 'lin $\perp$ lin' configuration (R_V and $R_\parallel$ in Fig. 5.7) the polarisation vectors are $e_V = \frac{1}{\sqrt{2}}\{1, 0, 1\}$ and $e_\parallel = \frac{1}{\sqrt{2}}\{1, 0, -1\}$, giving $\delta\phi = \pi$.

In deriving Eq. 4.12 we neglected the off-resonant terms and made a Lamb–Dicke approximation. As we will see in the next section, the higher order terms in the Lamb–Dicke approximation introduce a sensitivity to the initial state of the motion, and the off-resonant terms can lead to substantial errors for 'fast' gates, where the gate detuning δ is no longer small compared with ω_z.

4.4   Sources of Error

The error sources in a two-qubit gate can be split into two important classes. There are technical errors due to our imperfect control of experimental parameters (e.g. detunings or beam powers), and more fundamental errors due to the nature of the physical interaction (e.g. due to a finite Lamb–Dicke parameter, or Raman scattering). In this section we consider both of these classes of error in the abstract as they apply to a light-shift gate. In Chap. 8 we apply these to our experiment.

We define 'gate error' in the following as the infidelity in producing the maximally entangled state $|\psi_{\mathrm{id}}\rangle = \frac{1}{2}(|\downarrow\downarrow\rangle + i\,|\downarrow\uparrow\rangle + i\,|\uparrow\downarrow\rangle + |\uparrow\uparrow\rangle)$ from the separable state $|\psi_0\rangle = \frac{1}{2}(|\downarrow\downarrow\rangle + |\downarrow\uparrow\rangle + |\uparrow\downarrow\rangle + |\uparrow\uparrow\rangle)$. That is, $\epsilon = 1 - \langle\psi_{\mathrm{id}}|\,\rho\,|\psi_{\mathrm{id}}\rangle$, where $|\psi_{\mathrm{id}}\rangle$ is the desired maximally entangled output state, and ρ is the (in general) mixed state produced by application of the gate to $|\psi_0\rangle$. This definition matches the experimental work we perform in Chap. 8, and is the standard definition used in the field; however in a more general computational context the average gate error over all possible input states is more appropriate. For all the sources of error considered here, with the exception of photon scattering (Sect. 4.4.10) which is considered separately, the gate error after averaging over all pure input states is $4/5$ of the error for the input state $|\psi_0\rangle$.

In the following treatment we assume that we can estimate the total gate error by adding each of the (small) error contributions together. This is not in general true; for example, if we mis-set the gate detuning we can compensate somewhat by adjusting the gate Rabi frequency. Intuition suggests, and numerical simulations confirm, that all the incoherent error sources we consider accurately add. The coherent error sources do cross-couple, but for the typical magnitude and distribution of errors we expect in our experiment ignoring the cross-couplings is a reasonable approximation.

To analyse the effect of most of these error sources we numerically integrate the master equation in the Lindblad form. We typically model two spins coupling to one motional mode, and truncate the motional space to the lowest 8–20 states (depending on whether the motion is starting in the ground state or a thermal state). Integrating with the MATLAB solver 'ode45' and tolerance of 10^{-10} we find gate errors of $<10^{-6}$ with no error terms added to the simulation. This is sufficiently precise as our error sources are typically 10^{-5} or larger.

4.4.1 Mis-Set Gate Detuning

If the gate detuning is incorrectly set then the loops in phase space do not completely close, leading to the spin and motion remaining entangled at the end of the gate operation. We calculate this error by expanding the analytic solution (Eq. 4.10) for the fidelity about $\mathcal{F} = 1$. We find that the error is, for small detuning errors

$$\epsilon_\delta = \frac{1 + 2K(1 + 2\bar{n})}{16K^2} \left(\kappa t_g \right)^2 \tag{4.14}$$

where κ is the absolute error in the detuning from the ideal value, K is the number of loops of the gate, and $\bar{n}$ is the gate motional mode mean thermal occupation. If this error is significant there are composite pulse techniques that can be used to reduce the sensitivity [HCD+12].

4.4.2 Mis-Set Rabi Frequency

If the gate Rabi frequency is mis-set from the correct value the geometric phases accumulated will differ from the ideal values. We calculate this error by expanding the analytic solution (Eq. 4.10) for the fidelity about $\mathcal{F} = 1$. We find that the error is, for small Rabi frequency errors

$$\epsilon_\Omega = \frac{\pi^2}{4} \left(\frac{\delta\Omega}{\Omega_0} \right)^2 \tag{4.15}$$

where $\delta\Omega/\Omega_0$ is the fractional change in the Rabi frequency from the ideal value Ω_0.

4.4.3 Unequal Illumination

If the two ions are unequally illuminated by the Raman beams, the Raman Rabi frequency will be different for each ion. This means that, in general, each spin state will feel a different force amplitude, given by

$$\uparrow\uparrow : \quad (1+\alpha)\Omega_\uparrow - (1-\alpha)\Omega_\uparrow$$
$$\uparrow\downarrow : \quad (1+\alpha)\Omega_\uparrow + (1-\alpha)\Omega_\downarrow$$
$$\downarrow\uparrow : \quad -(1+\alpha)\Omega_\downarrow - (1-\alpha)\Omega_\uparrow$$
$$\downarrow\downarrow : \quad -(1+\alpha)\Omega_\downarrow + (1-\alpha)\Omega_\downarrow \qquad (4.16)$$

where the Raman Rabi frequency is scaled by $(1+\alpha)$ and $(1-\alpha)$ on the two ions respectively. If $\Omega_\uparrow = \Omega_\downarrow$ the force magnitude will be equal for $\uparrow\uparrow$ and $\downarrow\downarrow$, and also equal for $\uparrow\downarrow$ and $\downarrow\uparrow$. Thus by adjusting the Raman beam power a perfect entangling gate can implemented.

If, however, $\Omega_\uparrow \neq \Omega_\downarrow$ then each of the 4 spin states has a different force magnitude, and hence gains a different geometric phase. This gives an error in the gate operation that cannot be compensated for with a Raman beam power adjustment. This error is, to second order in α,

$$\epsilon_u = \alpha^2 \pi^2 \left(\frac{1-k}{1+k}\right)^2 \qquad (4.17)$$

where $k := \Omega_\uparrow / \Omega_\downarrow$.

This error can be removed by implementing a two-loop ($K = 2$) gate, with a qubit carrier π-pulse between the two loops. This makes that each qubit spends an equal amount of time in $\uparrow$ and $\downarrow$, effectively averaging k to 1, and hence removing the error. This technique is described further in Sect. 8.4.

4.4.4 Thermal Errors

As we have shown in Sect. 4.2 geometric phase gates are motionally insensitive in the Lamb–Dicke approximation. However typical experiments have $\eta \sim 0.1$ so neglecting terms $O(\eta^2)$ is not a particularly good approximation. Thermal occupation of both the motional mode used for the gate and the spectator motional mode(s) reduce the gate fidelity. We start by analysing the effect of a warm spectator mode.

Neglecting off-resonant effects and ambient heating, the motional distribution of the spectator mode does not change over the course of the gate. The effect of a spectator mode motional excitation is to cause the ion to explore a larger extent of the coupling field such that the amplitude of the driving field averaged over the motional wave-packet decreases. As we saw in Sect. 2.3, for an ion in motional state $|n\rangle$

$$\Omega_{n,n} := \Omega_{\text{free}} \left| \langle n | \exp \left(i\eta(a + a^\dagger)\right) | n \rangle \right|$$
$$= \Omega_{\text{free}} e^{-\eta^2/2} (1 - n\eta^2 + O(\eta^4)) \qquad (4.18)$$

We can see that the effect of spectator mode occupation is to change the spin-dependent force Rabi frequency as a function of n. If the spectator mode is in a specific Fock state this can be perfectly corrected by adjusting the Rabi frequency, but for a thermal state this will cause an uncorrectable error. Using the previously derived expression for the gate error from mis-set Rabi frequency (Eq. 4.15) we can express the gate error for a thermally occupied spectator mode, assuming the gate is perfect for $\bar{n} = 0$, as

$$
\begin{aligned}
\epsilon_{\bar{n}_s} &= \sum_n p(n, \bar{n}) \frac{\pi^2}{4} \left(\frac{\Omega_{n,n}}{\Omega_{0,0}} - 1 \right)^2 \\
&= \frac{\pi^2 \eta^4}{4} \sum_{n=0}^{\infty} \frac{\bar{n}^n}{(\bar{n} + 1)^{n+1}} n^2 \\
&= \frac{\pi^2 \eta^4}{4} \bar{n}(2\bar{n} + 1)
\end{aligned}
\tag{4.19}
$$

The result of this analytic approximation is plotted alongside the full numerical result in Fig. 4.1. Also plotted is the minimum gate error achievable after optimising the gate Rabi frequency for each temperature.

We now consider the effect of thermal occupation of the gate mode itself. This is more complex than for the spectator modes as the motional state evolves over the course of the gate – even for a gate mode initially cooled to the ground state there will still be a modification from the out-of-Lamb–Dicke-approximation terms. We do not attempt to model this analytically.

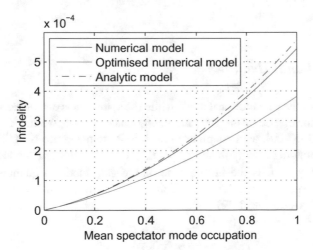

Fig. 4.1 Gate error from thermal occupation of the spectator mode for $\eta = 0.094$, corresponding to the two-ion breathing mode for a $\omega_z = 2\pi \cdot 1.93\,\text{MHz}$ $^{43}\text{Ca}^+$ crystal. The analytic model (*dashed line*) and the numeric result (*solid blue line*) are for a gate with Rabi frequency optimised for $\bar{n} = 0$. The 'optimised' numeric result (*solid green line*) is the minimum error achievable after empirically optimising the gate Rabi frequency for each temperature

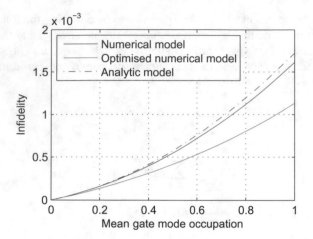

Fig. 4.2 Gate error from thermal occupation of the gate mode for $\eta = 0.124$, corresponding to the two-ion centre-of-mass mode for a $\omega_z = 2\pi \cdot 1.93\,\text{MHz}$ $^{43}\text{Ca}^+$ crystal. The *solid blue line* is the numeric result for a gate with Rabi frequency optimised for $\bar{n} = 0$. The *dashed line* is the analytic model derived for the *spectator* mode, which also appears to describe this error well. The 'optimised' numeric result (*solid green line*) is the minimum error achievable after optimising the gate Rabi frequency for each temperature

Our numeric simulation of the error from a thermally occupied gate mode (Fig. 4.2) shows that our *spectator* mode error model (Eq. 4.19) agrees well with these numerical results.

4.4.5 Motional Heating

Any heating of the gate motional mode while the spin state is entangled with motion, as it is over the course of the gate operation, leads to error. We model the heating, following [TMK+00], as a coupling to an infinite temperature bath. This leads to Lindblad operators $L_- = \sqrt{\gamma}a$ and $L_+ = L_-^\dagger$, where a is the gate mode annihilation operator and $\gamma = \dot{\bar{n}}$ is the heating rate of the gate mode in quanta-per-second. Integrating the master equation we find that the gate error ϵ_h from heating is described by (for small error, $\epsilon_h \ll 10\%$)

$$\epsilon_h = \frac{\gamma t_g}{2K} \tag{4.20}$$

where t_g is the gate duration, and K is the number of loops of the gate. As expected ϵ_h decreases with K, because the level of spin-motion entanglement decreases with K. In the large-K limit the motional excitation becomes completely virtual, and the mechanism is intensive to heating [SrM199].

4.4.6 Motional Dephasing

Any dephasing of the gate motional mode while the spin state is entangled with motion, as it is over the course of the gate operation, leads to error. We model the dephasing of the gate motional mode with the Lindblad operator $L = \sqrt{\frac{2}{\tau}}a^\dagger a$, with τ the 'motional coherence' time. This Lindblad operator causes the coherence between motional states $|n\rangle$ and $|m\rangle$ to decay at rate $(n - m)^2/\tau$. Integrating the master equation we find that the gate error from motional dephasing is described by

$$\epsilon_d = \alpha_K \frac{t_g}{\tau} \tag{4.21}$$

were $\alpha_K = \{0.686, 0.297, 0.137\}$ for $K = \{1, 2, 4\}$. As expected ϵ_d decreases with K, for the same reason as for the heating error decreases.

4.4.7 Amplitude and Phase Noise

Broadly speaking, there are two regimes of noise: 'drifts' much slower than the gate operation and broad-band noise much faster than the gate. A drift in the Raman beam phase does not affect the gate, and a drift in the Raman beam Rabi frequency can be modelled as a randomly mis-set Rabi frequency. To estimate the error from broad-band noise we assume the noise is white (i.e. the spectrum is flat), which is usually a reasonable approximation.

We wish to find the average dynamics of our system with a white noise term $\xi(t)H'(t)$ added to the Hamiltonian, where $\xi(t)$ is the noise process and $H'(t)$ is the part of the Hamiltonian affected by the noise. Following [Har13], we model the noise with the Lindblad operator $L = \sqrt{\Gamma}H'(t)$, where $\langle \xi(t)\,\xi(0)\rangle = \Gamma\delta(t)$, with $\delta(t)$ the Dirac delta function. For weak noise the dynamics are affected in the same way by both phase noise and amplitude noise: in either case we can write the system Hamiltonian as $H = (1 + \xi(t))\,H'(t)$.

Integrating the master equation with this Lindblad operator we find that the error from white amplitude noise on the Rabi frequency, or white noise on the Raman beam difference phase ϕ_0, is described by

$$\epsilon_n = \beta_K \frac{\Gamma}{t_g} \tag{4.22}$$

where $\beta_K = \{8.5, 13.3, 22.5\}$ for $K = \{1, 2, 4\}$, and Γ is the single-sided power spectral density in (fractional) power per Hz.[2] As a comparison, a single-qubit

---

[2]This is best defined operationally: If we have a 0 dBm signal of carrier frequency f on a white noise floor (amplitude or phase noise, but not both), we measure a noise power at frequency $f \pm \delta f$ of $10 \log_{10} \Gamma$ dBm/Hz.

π-pulse with duration t_g has $\beta = 2.5$: our two-qubit operations are at least a factor 4 more sensitive to noise than our single-qubit operations, as we might expect from the more complicated dynamics.

4.4.8 Motional Kerr Cross-Coupling

In the vast majority of trapped ion experiments, including this one, the trap potentials are accurately described as harmonic [HHJ+11]. However the form of the Coulomb repulsion between the ions leads to an inherent and unavoidable anharmonicity for any motional mode that involves relative motion of the ions.

For a two-ion crystal this leads to a coupling between the axial breathing mode and the radial rocking modes. Physically, the breathing mode frequency depends on the mean separation between the ions, and the radial rocking mode excitation changes this mean separation. This leads to a Kerr-type cross-coupling given by [RMK+08, NRJ09]

$$\mathcal{H} = \chi \, a^\dagger a \, b^\dagger b \tag{4.23}$$

$$\chi = -\omega_b \left(1 + \frac{\frac{1}{2}\omega_b^2}{4\omega_r^2 - \omega_b^2} \right) \left(\frac{\omega_z}{\omega_r} \right) \left(\frac{2\hbar\omega_z}{\alpha^2 mc^2} \right)^{1/3} \tag{4.24}$$

where $\omega_b = \sqrt{3}\omega_z$ is the breathing mode frequency, $\omega_r = \sqrt{\omega_\perp^2 - \omega_z^2}$ is the rocking mode frequency ($\omega_\perp$ is the radial trap frequency), and α is the fine-structure constant. This coupling leads to modulation of the axial breathing mode frequency by χn_r or vice versa. A thermal occupation of the 'other' mode leads to a fluctuating mode frequency and hence fluctuating gate detuning. The variance of these fluctuations for the breathing mode is

$$\langle \delta\omega_b^2 \rangle = \chi^2 \bar{n}_r (2\bar{n}_r + 1) \tag{4.25}$$

where $\bar{n}_r$ is the mean rocking mode occupation number. Using Eq. 4.14 we can calculate the resulting error for a gate performed on the axial breathing mode from the coupling to one of the rocking modes

$$\epsilon_\chi = \frac{1}{8K} (2\bar{n}_b + 1) \left(\chi t_g \right)^2 \bar{n}_r (2\bar{n}_r + 1) \tag{4.26}$$

As there are two independent radial modes, typically with similar frequencies, the total error from the cross-coupling will be roughly twice this. For our typical experimental trap frequencies (2 MHz axial, 4.5 MHz radial) the cross-coupling coefficient is $\chi = -2\pi \cdot 21$ Hz. This gives mode frequency fluctuations of $\langle \delta\omega_b^2 \rangle = 2\pi \cdot 210$ Hz for $\bar{n}_r = 7$. For a 100 μs two-loop gate the total error (from both rocking modes) is $\epsilon = 2 \times 10^{-3}$.

A similar Kerr-type coupling exists *between* the two radial rocking modes. Assuming the two radial modes have similar frequencies, the non-linear coefficient is

$$\chi = \left(\frac{2\hbar\omega_z}{\alpha^2 mc^2} \right)^{1/3} \frac{3}{4} \frac{\omega_z^3}{\omega_r^2 - \omega_z^2} \tag{4.27}$$

This does not affect gates perform on the axial breathing mode, but does lead to a significant additional source of error for gates performed on one of the radial rocking modes.

4.4.9 Spin Dephasing from Rayleigh Scattering

Rayleigh photon scattering from the Raman beams can lead to spin dephasing (Sect. 3.3.2); the Lindblad operator for this dephasing is given by Eq. 3.8. Integrating the master equation we find the error from this spin dephasing to be described by, for small error

$$\epsilon_{\text{Ray}} = \frac{1}{2} \Gamma_{\text{Rayleigh}} t \tag{4.28}$$

4.4.10 Raman Photon Scattering

If the qubit is 'closed' under scattering, and the scattering rate out of both qubit states is the same, we can model the Raman photon scattering process as a spin-flip error on our qubit, with Lindblad operators $L = \sqrt{\Gamma_{\text{Raman}}} \sigma_+$ and conjugate for each qubit (where Γ_{Raman} is the scattering rate for a single ion from the pair of Raman beams needed to implement the gate). In general these are reasonable, but not accurate, assumptions. Integrating the master equation with this Lindblad operator, we find the error from the Raman photon scattering is described by, for small error,

$$\epsilon_{\text{Raman}} = \frac{3}{2} \Gamma_{\text{Raman}} t \tag{4.29}$$

Averaging over a random uniform selection of pure initial states we find the mean error is 7% larger than this result.

In the opposite limit to a 'closed' qubit, any scattering event removes the ion from the qubit sub-space. If either of the two qubits suffers a Raman scattering event, the resulting state fidelity is 0. Thus the error is given by

$$\epsilon_{\text{Raman}} = 2\Gamma_{\text{Raman}} t \tag{4.30}$$

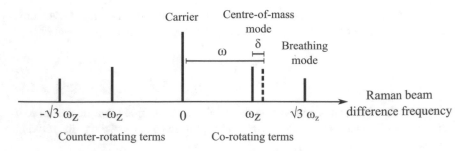

Fig. 4.3 Relevant frequencies to the light-shift gate. The Raman beam difference frequency ω is detuned by δ from the centre-of-mass mode. The carrier light-shift term is off-resonant by $\omega_z + \delta$, and the co-rotating breathing mode term by $(\sqrt{3} - 1)\omega_z - \delta$

4.4.11 Off-Resonant Excitation

When we expanded the full system Hamiltonian (Eq. 4.11) we dropped a number of off-resonant terms (Fig. 4.3). In this section we calculate the errors caused by the presence of these terms. We assume that the gate speed is low enough that the off-resonant terms make only a small modification to the dynamics.

For two ions in the Lamb–Dicke regime there are three possible off-resonant effects that can reduce the gate fidelity. The first is the 'carrier' light-shift, off-resonant by $\approx \omega_z$ (assuming we are performing the gate using the centre-of-mass mode). This terms adds a phase shift onto the qubits that depends on the (uncontrolled) initial optical difference phase of the Raman beams, ϕ_0, reducing the average gate fidelity. The second effect is excitation of the 'other' axial motional mode of the ions, the breathing mode – the spin-dependent force is off-resonant with this mode by $\approx (\sqrt{3} - 1)\omega_z$ or $\approx (\sqrt{3} + 1)\omega_z$ (for the co-rotating and counter-rotating terms respectively). The third effect is the counter-rotating term of the spin-dependent force itself at $\approx 2\omega_z$.

We expect the magnitude of the errors to scale as Ω^2/δ'^2, where δ' is the detuning of the term under consideration from resonance. This means that the co-rotating terms will dominate and we can ignore the counter-rotating terms when estimating the error. The breathing mode excitation and the carrier excitation have similar detunings $\sim \omega_z$, so we expect the breathing mode excitation to be roughly $\eta_b^2 \sim 10^{-2}$ times smaller than the carrier excitation. In the following we find that significant errors can occur if we use rectangular pulses, but that if we shape the turn-on and turn-off transients of the Rabi frequency to have time-scale $1/\delta$ we can suppress these errors substantially. We calculate these off-resonant errors using an approach similar to [SIHL14].

4.4.11.1 Light-Shift Error

The carrier light-shift term for two ions with motional phase difference $\phi_m = \pi$ (as for two ions separated by a half-integer number of standing wavelengths), dropping a global phase, is

$$\mathcal{H}_c = \frac{1}{2} \sum_j \bar{\Omega} \sigma_{z,j} e^{i\phi_0} e^{-i\omega t} (-1)^j + \text{h.c.} \tag{4.31}$$

where $\bar{\Omega} := \frac{1}{2}(\Omega_\uparrow + \Omega_\downarrow)$, and ω is the Raman beam difference frequency. As this Hamiltonian commutes with the dominant spin-dependent force part of the Hamiltonian we can integrate the Hamiltonians separately and multiply the propagators. Doing this we find the propagator U_c to be

$$U_c = U_{z,1}(\phi_1) U_{z,2}(\phi_2) \tag{4.32}$$

$$U_{z,j}(\phi) = \exp(i\sigma_{z,j}\phi/2) \tag{4.33}$$

$$\phi_j = e^{i\phi_0} \theta_j - e^{-i\phi_0} \theta_j^* \tag{4.34}$$

$$\theta_j = -i(-1)^j \int_0^{t_g} \bar{\Omega}(t) e^{-i\omega t} dt \tag{4.35}$$

We now want to find the infidelity caused by these phase shifts. We assume that we have carefully adjusted the polarisations of the two Raman beams used in the gate to null the single-beam light-shifts. Thus the light-shift errors will come from the oscillating interference pattern between the two beams. Considering applying the propagator of Eq. 4.35 to our standard initial spin-state $|\psi\rangle = R(\pi/2, 0)^{\otimes 2} |\downarrow\downarrow\rangle$ we find

$$\mathcal{F} = |\langle\psi| U_c |\psi\rangle|^2 \tag{4.36}$$

$$= \cos^2 \frac{\phi_1}{2} \cos^2 \frac{\phi_2}{2} \tag{4.37}$$

$$\approx 1 - \frac{1}{4}(\phi_1^2 + \phi_2^2) \tag{4.38}$$

where we have assumed that the accumulated carrier phase shift is small, and hence made a small-angle approximation. Averaging over a uniformly distributed Raman beam initial phase ϕ_0 we find

$$\langle\phi_j^2\rangle = 2|\theta_j|^2 \tag{4.39}$$

$$\bar{\epsilon}_c = |\theta|^2 \tag{4.40}$$

as $|\theta_1| = |\theta_2|$. For a rectangular gate Rabi frequency pulse with K loops we find

$$\epsilon_c = 4K \left(\frac{\pi}{\eta_c \omega t_g}\right)^2 \sin^2 \frac{\omega T}{2} \tag{4.41}$$

In the worst case (the sin term at a maximum) a 100 motional-mode-period 2 loop gate with $\eta_c = 0.12$ (a ~50 μs gate on 2 ^{43}Ca$^+$ ions in a $\omega_z \sim 2\pi \cdot 2$ MHz trap) has error $\epsilon_c = 1.4\%$. This is a substantial error! The peak single-ion phase shift accumulated for these parameters, for the worst possible initial Raman beam difference phase and gate length, is $\phi = 14°$.

For technical reasons, in our experimental implementation we perform a two-loop gate ($K = 2$), with a carrier π-pulse between the two loops. This modifies the carrier excitation error somewhat – for the worst possible phase difference between the two gate pulses the error is ≈ 2 times larger than that given by Eq. 4.41.

Shaping the gate pulse Rabi frequency with an envelope of duration t_{shape} dramatically reduces this error. Figure 4.4 shows the ratio of the gate error with the pulse-shaping to that with a rectangular pulse, versus the pulse-shaping duration, for both a Blackman pulse shape and a sin^2 pulse shape (shape function sin$^2 \frac{\pi t}{2t_{\text{shape}}}$). These pulse-shape functions are plotted in Fig. 4.5. Importantly for experimental work, the gate error is not sensitive to the precise pulse shape. A $t_{\text{shape}} = 1$ mode-period pulse-envelope length gives a reduction in error of ≈ 10 (i.e. multiplies Eq. 4.41 by 10^{-1}), and $t_{\text{shape}} = 3$ mode-periods (~ 1.5 μs for a 2 MHz trap) reduces the error by $\approx 10^3$, giving an error $\epsilon_c = 10^{-5}$ for a 100 period gate.

4.4.11.2 Breathing Mode Excitation

We now consider excitation of the breathing mode. The Hamiltonian for this term (dropping the counter-rotating term) is

$$\mathcal{H}_{\text{breath}} = \frac{i\eta_b}{2} \left(2\Omega_\uparrow |\uparrow\uparrow\rangle \langle\uparrow\uparrow| - 2\Omega_\downarrow |\downarrow\downarrow\rangle \langle\downarrow\downarrow| \right) e^{i\phi_0} e^{-i\omega t} a_b^\dagger e^{+i\omega_b t} + \text{h.c.} \quad (4.42)$$

In the following we assume $\Omega_\uparrow \approx \Omega_\downarrow$. The effect of this Hamiltonian is to displace the spin states in the breathing mode phase-space and to add a geometric phase.

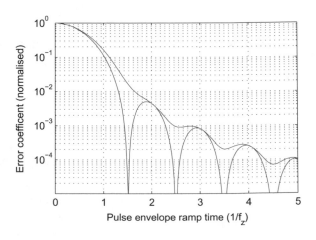

Fig. 4.4 Ratio of off-resonant excitation for a shaped gate pulse versus a rectangular gate pulse. The *blue line* is for a Blackman pulse-shape and the *red line* for a sin^2 pulse-shape

Fig. 4.5 The three
pulse-shapes we consider; a
rectangular pulse (*black*), a
Blackman windowed pulse
(*blue*), and a $\sin^2$ shaped
pulse (*red*). The time axis is
normalised so that the
pulse-shape envelope length,
t_{shape}, is 1

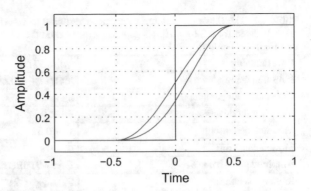

The principal source of error is from the displacement. Integrating this Hamiltonian
ignoring the geometric phase we find it applies a displacement $D(\alpha)$ to $\uparrow\uparrow$ and
$D(-\alpha)$ to $\downarrow\downarrow$, where we define α shortly. The error of this propagator compared
with the identity is, for small displacements,

$$\epsilon_b \approx \frac{1}{2}|\alpha|^2 (2\bar{n}_b + 1) \tag{4.43}$$

If we assume the gate is slower than a few periods of motion the counter-rotating
term $((\sqrt{3} + 1)\omega_z)$ is negligible compared to the co-rotating term $((\sqrt{3} - 1)\omega_z)$,
hence there is no dependence on the Raman beam phase ϕ_0 and the displacement is
given by

$$\alpha = \int_0^{t_g} \eta\Omega e^{-i(\omega - \omega_b)t}\,\mathrm{d}t \tag{4.44}$$

For a square pulse and cold motional mode

$$\epsilon_b = 2K \left(\frac{\eta_b \pi}{\eta_c t_g (\omega - \omega_b)} \right)^2 \sin^2 \frac{(\omega - \omega_b)t_g}{2} \tag{4.45}$$

A 100 mode-period 2 loop gate has error $\epsilon_b = 0.01\%$ from this term without pulse-
shaping. With pulse-shaping this error will reduced in the same was as the carrier
excitation error. Thus this error will always be negligible compared to the carrier
error ϵ_c.

4.5 Error Summary

How well do we need to control each of the potential sources of error we have
discussed? Table 4.1 lists how large each of these error sources has to be to contribute
a 10^{-4} error for a typical $100\,\mu\text{s}$ duration gate. We note that there are many other
sources of experimental error – these are discussed in detail in Chap. 8.

Table 4.1 The limits on each error parameter required to contribute a $<10^{-4}$ gate error. These are calculated for a 2-loop $100\,\mu s$ gate implemented on the $1.93\,MHz$ centre-of-mass mode of a two-ion $^{43}Ca^+$ crystal

Gate detuning	$|\kappa| < 57\,Hz$
Rabi frequency	$|\delta\Omega/\Omega_0| < 0.6\%$
Gate mode temperature	$\bar{n}_c < 0.14$
Spectator mode temperature	$\bar{n}_b < 0.32$
Motional dephasing	$\tau > 0.3\,s$
Motional heating	$\dot{\bar{n}} < 4\,s^{-1}$
Amplitude/phase noise	$\Gamma < -91\,dBc/Hz$
Rayleigh scattering	$\Gamma_{Rayleigh} < 2\,s^{-1}$
Raman scattering	$\Gamma_{Raman} < 0.7\,s^{-1}$
Off-resonant excitation	$t_{shape} > 1\,\mu s$

References

[BSPR12] Bermudez, A., P.O. Schmidt, M.B. Plenio, and A. Retzker. 2012. Robust trapped-ion quantum logic gates by continuous dynamical decoupling. *Physical Review A* 85 (4): 1–5.

[Ber84] Berry, M.V. 1984. Quantal Phase Factors Accompanying Adiabatic Changes. *Proceedings of the Royal Society A: Mathematical, Physical and Engineering Sciences* 392 (1802): 45–57.

[Car65] Carruthers, P. 1965. Coherent States and the Forced Quantum Oscillator. *American Journal of Physics* 33 (7): 537.

[CZ95] Cirac, J.I., and P. Zoller. 1995. Quantum Computations with Cold Trapped Ions. *Physical Review Letters* 74 (20): 4091.

[GRZC03] García-Ripoll, J.J., P. Zoller, and J.I. Cirac. 2003. Speed Optimized Two-Qubit Gates with Laser Coherent Control Techniques for Ion Trap Quantum Computing. *Physical Review Letters* 91 (15): 157901.

[Har13] Harty, T.P. 2013. High-Fidelity Microwave-Driven Quantum Logic in Intermediate-Field 43Ca+. Ph.D. thesis.

[HCD+12] Hayes, D.L., S. Clark, S. Debnath, D. Hucul, I. Inlek, K. Lee, Q. Quraishi, and C. Monroe. 2012. Coherent Error Suppression in Multiqubit Entangling Gates. *Physical Review Letters* 109 (2): 020503.

[HML+06] Home, J.P., M.J. McDonnell, D.M. Lucas, G. Imreh, B.C. Keitch, D.J. Szwer, N.R. Thomas, S.C. Webster, D.N. Stacey, and A.M. Steane. 2006. Deterministic entanglement and tomography of ion-spin qubits. *New Journal of Physics* 8 (9): 188–188.

[HHJ+11] Home, J.P., D. Hanneke, J.D. Jost, D. Leibfried, and D.J. Wineland. 2011. Normal modes of trapped ions in the presence of anharmonic trap potentials. *New Journal of Physics* 13 (7): 073026.

[JPK00] Jonathan, D., M.B. Plenio, and P.L. Knight. 2000. Fast quantum gates for cold trapped ions. *Physical Review A* 62 (March): 1–10.

[KAN+14] Kotler, S., N. Akerman, N. Navon, Y. Glickman, and R. Ozeri. 2014. Measurement of the magnetic interaction between two bound electrons of two separate ions. *Nature* 509 (7505): 376–380.

[LBD+05] Lee, P., K.-A. Brickman, L. Deslauriers, P.C. Haljan, L.-M. Duan, and C. Monroe. 2005. Phase control of trapped ion quantum gates. *Journal of Optics B: Quantum and Semiclassical Optics* 7 (10): S371–S383.

[LDM+03] Leibfried, D., B. DeMarco, V. Meyer, D.M. Lucas, M.D. Barrett, J.W. Britton, W.M. Itano, B. Jelenković, C.E. Langer, T. Rosenband, and D.J. Wineland. 2003. Experimental demonstration of a robust, high-fidelity geometric two ion-qubit phase gate. *Nature* 422 (6930): 412–415.

[Mag54] Magnus, W. 1954. On the exponential solution of differential equations for a linear operator. *Communications on Pure and Applied Mathematics* VII: 649–673.

[NRJ09] Nie, X.R., C.F. Roos, and D.F.V. James. 2009. Theory of cross phase modulation for the vibrational modes of trapped ions. *Physics Letters A* 373 (4): 422–425.

[Oze11] Ozeri, R. 2011. The trapped-ion qubit tool box. *Contemporary Physics* 52: 531–550.

[RMK+08] Roos, C.F., T. Monz, K. Kim, M. Riebe, H. Häffner, D.F.V. James, and R. Blatt. 2008. Nonlinear coupling of continuous variables at the single quantum level. *Physical Review A* 77 (4): 040302.

[vS03] Šašura, M., and A.M. Steane. 2003. Fast quantum logic by selective displacement of hot trapped ions. *Physical Review A* 67 (6): 062318.

[SrM199] Sørensen, A., and K. Mø lmer. 1999. Quantum Computation with Ions in Thermal Motion. *Physical Review Letters* 82 (9): 1971–1974.

[SIHL14] Steane, A.M., G. Imreh, J.P. Home, and D. Leibfried. 2014. Pulsed force sequences for fast phase-insensitive quantum gates in trapped ions. *New Journal of Physics* 16 (5): 053049.

[TMK+00] Turchette, Q.A., C.J. Myatt, B.E. King, C.A. Sackett, D. Kielpinski, W.M. Itano, C. Monroe, and D.J. Wineland. 2000. Decoherence and decay of motional quantum states of a trapped atom coupled to engineered reservoirs. *Physical Review A* 62 (5): 1–22.

Chapter 5
Apparatus

This chapter describes the apparatus used for the experiments in this thesis. There is a somewhat arbitrary separation between this chapter and Chap. 6 'Experiment Characterization'; construction and technical details are given in this chapter, and performance measurements are in Chap. 6. Figure 5.1 is a block-diagram of the experimental apparatus.

5.1 The Ion Trap

In this section we introduce the physical implementation of the ion trap and the vacuum system it resides in.

5.1.1 Trap Geometry

The ion trap is an 'Innsbruck style' blade trap (Fig. 5.2) [Gul03]. We drive one pair of blades with RF, ground the other pair, and apply DC to the end-caps. Using blades rather than (closer to) hyperbolic electrodes improves optical access without greatly reducing the geometry factor. The stainless steel blades and end-caps are supported by a pair of Macor ceramic 'crosses'. Three compensation electrodes, used to trim out residual electric fields at the centre of the trap, are also attached to the Macor crosses. They are positioned as close as possible to the trap centre without obstructing the optical access. The characteristic dimensions of the trap are given in Table 5.1.

The geometry factors are calculated by determining the fields from the electrodes over a small ($0.1 \times 0.1 \times 0.1$ mm) test volume at the centre of the trap using a Bound-

© Springer International Publishing AG 2017
C.J. Ballance, *High-Fidelity Quantum Logic in Ca⁺*,
Springer Theses, DOI 10.1007/978-3-319-68216-7_5

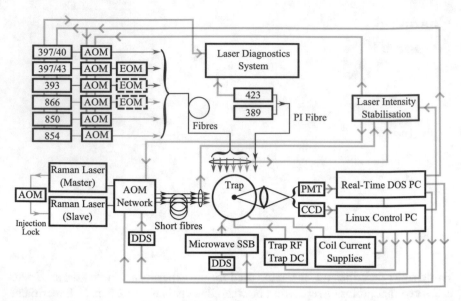

Fig. 5.1 A block-diagram of the experimental apparatus. Each of these blocks is described in this chapter

Table 5.1 Trap dimensions and geometrical factors α. The ion-blade separation is ρ_0, and the ion-endcap separation is z_0

ρ_0	0.5 mm
z_0	1.0 mm
α_x	0.495
α_y	0.460
α_z	0.285

ary Element Method solver.[1] Due to the linearity of Maxwell's equations we can superimpose the potential generated from the driven blades (with the endcaps and non-driven blades grounded) with the potential from the end-caps (with all the blades grounded). We fit the potentials obtained to the model of Eq. 2.4 to determine the geometry coefficients (Table 5.1). We note the asymmetry in the radial geometry coefficients ($\alpha_x \neq \alpha_y$) – this is due to trapping and anti-trapping from the axial RF potential caused by the asymmetric driving of the blades (see Sect. 2.2).

---

[1] 'Charged Particle Optics' (CPO).

Fig. 5.2 Rendered model of the ion trap. The *blue* electrodes are the blades, and the *pink* electrodes are the end-caps. The compensation electrodes are not shown

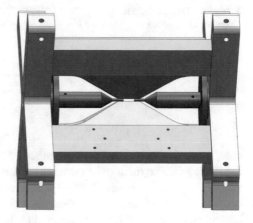

5.1.2 Vacuum System

The trap is mounted in a 'spherical octagon' vacuum chamber.[2] This provides 8 equally spaced CF40 flanges around the sides, and CF100 flanges on the top and bottom. The top CF100 flange has an AR coated view-port[3] principally used for imaging. The bottom CF100 flange has a custom baseplate with various electrical feed-throughs. View-ports are mounted on 6 of the CF40 flanges for laser access. Of these view-ports, 5 are directly mounted on the octagon, and one is mounted via a T-piece, to which a getter pump is connected.[4] The two CF40 ports along the trap axis (hence with no view of the trap centre due to the end-caps) do not have view-ports – one has the ion pump,[5] ion gauge,[6] and pump-out value, the other has the feed-through for the trap RF. The trap RF feed-through[7] has two pins, one connected to each driven trap blade.

The baseplate, to which the trap is mounted, contains electrical feed-throughs for the ovens, the trap DC connections, and the grounded trap-blade connections. It also has a recessed view-port from which the photo-ionisation beams and the vertical micro-motion detection beam exit (see Fig. 5.4). The main DC feed-through[8] connects to the 'reserve' oven and the 5 trap DC connections (2 end-caps and 3 compensation electrodes). Both ends of the grounded trap blades are connected to feed-throughs - this allows a choice of grounding options when driving microwave currents into the blades (Sect. 5.6). One of the blade feed-throughs has 2 pins,[9] the

---

[2]Kimball spherical octagon MCF600-SO20080.

[3]Torr Scientific VPZ100QBBAR-LN.

[4]SAES Sorb-AC GP 50, C50-ST101 cartridge.

[5]Varian VacIon Plus 20 (20L/s).

[6]Varian UHV-24P.

[7]FHP5-180C2-40C, 5kV, 180A.

[8]FHP1-C8-W, 8 pins, 1kV, 25A.

[9]FHP-C2-W, 500V, 15A.

Table 5.2 Trap RF helical
resonator dimensions

D	98.5 mm
d	45 mm
d_0	6.4 mm
b	90 mm
B	133 mm
τ	12 mm

other has 4 pins[10]; the main oven is connected to the remaining two pins of this second feed-through. The main oven contains an isotopically enriched $^{43}Ca^+ - {}^{40}Ca^+$ mixture (12% $^{43}Ca^+$, remainder $^{40}Ca^+$). The reserve oven contains natural abundance calcium (0.135% $^{43}Ca^+$). The ovens are resistively heated stainless-steel tubes containing granulated calcium, constructed to the group's standard design [All11].

The vacuum system is mounted on three aluminium brackets such that the focal point of the chamber (and hence the centre of the trap) is 100 mm above the surface of the optical table. The ion gauge reads a pressure of 1.1×10^{-11} Torr, which is close to the specified X-ray limit of the gauge system.

5.1.3 Trap RF Source

To achieve the desired radial trap frequencies we need $V_{RF} \sim 1kV$. To produce this we use a helical resonator [SSWH12, ZB61] to transform a high current, low voltage source into a high voltage, low current source. A helical resonator is similar to a $\lambda/4$ coaxial resonator, but the helical inner conductor dramatically shortens the length of the helical resonator – a 30 MHz coaxial resonator would be 2.5 m long compared with $\sim$10 cm for a helical resonator.

The dimensions of our helical resonator are given in Table 5.2, using the notation of Siverns [SSWH12]. The helical resonator consists of a copper cylinder (diameter D, length B) containing a copper helix (diameter d, length b, pitch τ, wire diameter d_0). One end of the helix is grounded (the low impedance end), and the other (high impedance) end is the high voltage output. The RF power is inductively coupled in to the resonator by a small coil (16 turns, 2.8 mm pitch, 25 mm diameter) co-axial with the main helix at the low impedance end. This input is matched to 50 Ω by tuning the position of the coupling coil relative to the helix.

With the high impedance end of the helix floating we measure a quality factor of $Q = 900$ and a resonant frequency of $f_0 = 48.8$ MHz. The RF output of the helical resonator connects directly to the trap RF feed-through via short ($\sim$2 cm) wires. The helical ground is connected to the grounded blades (on the bottom flange) via a copper ribbon ($\sim$1 cm width) – this grounding is critical. With the resonator attached to the trap we find $Q = 400$, $f_0 = 29.8$ MHz. The return loss is 35 dB (VSWR

---

[10]FHP-C4-W, 500V, 15A.

1.04 : 1). The voltage step-up is $\sim$39. If we disconnect the grounding ribbon (and hence increase the ground impedance) the quality factor decreases to $Q = 110$.

The trap RF chain consists of an RF synthesizer driving a digitally controlled attenuator,[11] then a power amplifier.[12] The attenuator allows the experimental control computer to adjust the RF amplitude with 31 dB dynamic range in 0.5 dB steps. The typical input power to the helical resonator is 4.6 W, which gives $V_{RF} \approx 830$ V (zero-peak) on the RF blades – this equates to a radial trap frequency of 4.3 MHz for $f_z = 2$ MHz.

5.1.4 Trap DC Source

The DC voltages for the end-caps and the three compensation electrodes are generated by a custom DAC (schematics in Appendix B.2). This five channel (16 bit) DAC produces output voltages -240 V $<$ V $< +240$ V with a resolution of 7 mV. The measured output noise is $\approx$1 mV rms in $0-20$ MHz suggesting that around 2 MHz the voltage spectral noise density is $\ll 1uV/\sqrt{Hz}$. At the normal end-cap voltage of 110V ($f_z \approx 2$ MHz) the measured output voltage drift over 2 hours is $\approx$0.6 mV – this corresponds to a worst-case secular frequency drift of 5Hz. The multimeter used to measure the voltage drift[13] has a specified accuracy of $\pm$2.8 mV and a 10 min transfer accuracy of 0.3 mV. As the drifts we measure over hours are of the order of the 10 min transfer accuracy it is unclear how much of the measured drift is from the DAC and how much from the multimeter, thus the measured drift is a upper limit. The DAC is controlled via an isolated SPI bus from the experimental control computer.

The voltages supplied by the DAC are fed to the trap electrodes via an additional filter mounted on the air side of the vacuum feed-through. The filter consists of a CRC π network followed by an LC network, with $C = 100$ nF, $R = 100 k\Omega$, and $L = 22$ uH. The first section has a pole at 16 Hz to take out 'slow' noise up to $\sim$ MHz. The second section has a pole at 110 kHz and provides good attenuation up to $\sim$100 MHz. For one of the end-cap channels and one of the compensation channels a 100nF capacitor after the filter couples a 'tickle' signal onto the electrodes. This 'tickle input' is used to bypass the filters when applying a weak oscillating voltage to excite the secular motion in order to determine the trap frequencies. A voltage spectral noise density of $1uV/\sqrt{Hz}$ at the motional mode frequency of 2 MHz applied to one end-cap produces a heating rate of $\dot{\bar{n}} = 1/s$. As the filter has a voltage attenuation $\sim$10^{-5} at the motional mode frequencies it is clear that the electronic noise is many orders of magnitude below that which would cause a heating rate of $1/s$.

---

[11]Minicircuits ZX76-31R5-PP-S+.

[12]Frankonia FLL-25. 1 dB compression at $P_O = 20$ W.

[13]Agilent 34401A.

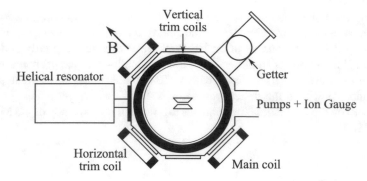

Fig. 5.3 Magnetic field coil layout. The pair of main coils and the pair of vertical coils are each connected in series

Table 5.3 Magnetic field coils. N is the total number of turns, R the mean coil radius, and l the distance from the coil centre to the ion. The final number is the expected field (Gauss) at the ion per amp of current

Coil	N	$2R$ (mm)	l (mm)	G/A
Main	2×171	95	120	2.3
Servo	2×129	95	120	1.7
Vertical trim	2×30	165	48	3.0
Horizontal trim	92	83.5	115	0.55

5.2 Magnetic Field Coils

There are 3 orthogonal sets of field coils around the vacuum system, described in Fig. 5.3 and Table 5.3. The main coils provide the bulk of the field, the trim coils are used to align the magnetic field precisely along the desired direction. For the standard magnetic field of 1.95 G the coil currents are $\{I_{main}, I_{vert}, I_{horiz}\} = \{1.1, 0.17, 0.03\}$A. The coils are driven by power supplies[14] remotely controlled by the experimental computer.

The main coils have an additional set of windings for the magnetic field 'servo', driven by a programmable current source (0–50 mA, 12 μA resolution). This gives a trim range of 20 mG with 20 μG resolution. A routine on the experimental control computer adjusts the current through this servo coil to keep the magnetic field at the ions constant in the presence of varying laboratory fields. We calculate the magnetic field error by sampling one point ($N = 200$ shots) on either side of a 100 μs π-pulse resonance on the ^{43}Ca$^+$ stretch qubit. This gives a lock noise of 60 μG (140 Hz) rms.

---

[14]Thurlby TTI QL series.

5.3 Imaging System

We collect and image the ions' fluorescence via an optical system suspended over the top view-port. The $NA = 0.25$ objective lens[15] images the ions onto either a photo-multiplier tube (PMT) or a camera. In all of the experiments described in this thesis we detect the state of the ions using the PMT, and only use the camera for diagnostic purposes. We select between the camera and PMT using a moveable beam-splitter. With the beam-splitter 'out' all the light is directed to the PMT. With the beam-splitter 'in' 65% of the light goes to the camera, and the rest to the PMT. A schematic of the imaging system can be found in [LAS+12].

The PMT[16] has a specified quantum detection efficiency of 16% at 397 nm. To minimise background counts (from room light and laser beam scatter) a 200 μm pinhole[17] is placed in the focal plane in front of the PMT. The magnification of the imaging system is $M = 5.9$. We determine the absolute collection efficiency of the imaging system and PMT (the probability of getting a 'click' from the PMT per photon emitted by the ions) to be $\eta = 0.282(7)\%$.

The Electron-Multiplying CCD (EM-CCD) camera[18] beam path has two extra elements after the beam-splitter; a lens with $f = -50$ mm positioned so as to give an extra magnification of ∼3, along with a coloured glass filter[19] to remove IR scatter. The addition of the lens increases the total magnification from ion to camera to $M = 19.5$. The pixel size of the camera is 16×16 μm. For a typical ion spacing of 3.5 μm (a two-ion $f_z = 2$ MHz crystal) the ion separation on the camera is 4.3 px, which is sufficient to resolve them.

5.4 Laser Systems

In this section we describe the photo-ionisation (PI), Doppler cooling, and readout laser sources and beam-paths, leaving discussion of the Raman lasers until Sect. 5.5.

5.4.1 Ca⁺ Lasers

The laser wavelengths needed for Doppler cooling and readout of Ca^+ are 393, 397, 850, 854 and 866 nm. The laser systems are designed to be able to cool and read-out $^{40}Ca^+$ and $^{43}Ca^+$ at the same time, as used in the experiments of Sect. 8.4. To do this

---

[15] Nikon ED PLAN 1.5x SM2-U.

[16] Hamamatsu H6180-01.

[17] Comar 200 HL 25.

[18] Andor iXon DU-897E.

[19] Melles-Griot FCG 433, BG-38 glass, T∼90% at 397 nm.

we use two separate 397 nm lasers, the '397/40' and the '397/43', and add EOMs on the 866 and 393 nm beams to span the isotope splittings.

All of the lasers are extended cavity diode lasers[20] (ECDLs) with optical isolators. The lasers are frequency stabilised by locking to piezo-tuned reference etalons[21] with Pound-Drever-Hall (PDH) locks. The frequencies of the lasers are tuned by the experimental control system DACs by varying the reference etalon piezo voltage. The phase modulation required for the lock is generated by modulating the laser diode current (for the 850, 866, and 397/43 lasers) or by an EOM in the lock path (for the 393 and 397/40).

Each laser is switched and amplitude-modulated with a double-pass acousto-optic modulator (AOM), then coupled into a single-mode polarisation-maintaining optical fibre to turn any beam pointing fluctuations in the laser source into easily measured and corrected power fluctuations. We measure typical on/off optical power ratios after the fibre of 10^{-6}. This is typically limited by the AOM RF source extinction. A sample of each laser beam is fed to a diagnostics system [All11] (the ARSES) to monitor the laser frequency and output spectrum. This consists of an 8 channel fibre switcher connecting each laser in turn to a wavemeter[22] and a UV/IR optical spectrum analyser. To remove amplified spontaneous emission that could excite spectator transitions during state-selective shelving or readout, the 866, 850, and 397 beams are filtered by blazed diffraction gratings.

The 397/43 laser needs to pump efficiently out of both the $S^3_{1/2}$ and $S^4_{1/2}$ manifolds. This is achieved by adding sidebands to the beam with a 3.2 GHz EOM.[23] The carrier of the laser is tuned near the $S^3_{1/2} \leftrightarrow P^4_{1/2}$ transition, while the sideband produced by the EOM addresses the $S^4_{1/2} \leftrightarrow P^4_{1/2}$ transition. The exact EOM frequency and sideband power are chosen to optimise the ion fluorescence, and are described in Sect. 6.2.1.

For simultaneous Doppler cooling and read-out of ^{43}Ca$^+$ and ^{40}Ca$^+$ we use EOMs on the 866 and 393 at 3.4 and 1.94 GHz respectively. The 866 carrier is tuned to address ^{43}Ca$^+$, and the blue sideband used to repump ^{40}Ca$^+$. The 393 carrier is tuned to the ^{43}Ca$^+$ $S^4_{1/2}$-$P^5_{3/2}$ transition, and the red sideband to the ^{40}Ca$^+$ transition. Further details on this dual-species readout scheme can be found in [Szw09, Lin12].

5.4.2   Photo-Ionisation Lasers

The photo-ionisation (PI) lasers, at ~389 and 423 nm, have much looser requirements than the other lasers. The 423 nm laser excites a dipole transition with a linewidth of 35 MHz, and consists of an ECDL with an optical isolator. As we reload infrequently we do not lock this laser, and instead rely on manual tuning and the passive stability of the laser frequency. The 389 nm laser excites a transition to the continuum, so the

---

[20]Toptica DL-100.

[21]NPL 'Low Drift Etalon', 1.5 GHz free spectral range, specified drift < 0.5MHz/hour.

[22]High Finesse WS-7, 10 MHz resolution.

[23]New Focus 4431M.

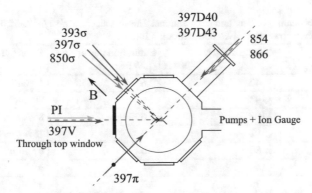

Fig. 5.4 Geometry of the laser beams illuminating the ions (excluding Raman beams). The main Doppler cooling beams enter from the *top-right*. There are three σ polarised optical pumping beams, and one π polarised beam. The 'vertical' 397 beam (for micro-motion compensation) and the photo-ionisation (PI) beams enter at 30° to the vertical through the *top view-port*

laser wavelength is unimportant (to a few nm). Thus we use a bare diode (rather than an ECDL) without an optical isolator. As the PI beams do not need to be switched on or off fast they are switched with a mechanical shutter. The two PI beams are combined on a dichroic mirror and coupled into a single-mode fibre.

5.4.3 Beam-Paths

Figure 5.4 is a sketch of the geometry of the laser beams illuminating the ions (excluding the Raman beams). Table 5.4 gives the beam spot sizes and polarisations. All of the beams (except the PI beams) are intensity-stabilised. The output beam from each of the fibres carrying light to the trap passes through a PBS to turn any polarisation variations from the fibre into power variations. A sample of the beam is then fed onto a photo-diode connected to the intensity stabilisation 'noise-eater' (Sect. 5.7.2).

The fibre output beam from the 397/43 and the 397/40 lasers is switched by an AOM network into several different paths. The beams can be switched into the Doppler-cooling paths (the $397D40$ and $397D43$ paths), the σ optical pumping paths (the $397\sigma40$ and $397\sigma43$ paths), or the π optical pumping paths (the $397\pi40$ and $397\pi43$ paths).

The PI beams and 397 nm vertical micro-motion compensation beams are combined on a PBS and enter via the top (imaging) window. The Doppler cooling and repumping beams enter from the top-right in Fig. 5.4. The IR beams and the UV beams are combined through a UV mirror, and are jointly steered by a dual-band (IR and UV) mirror before the focussing lens. The two IR beams (854 and 866) are combined on a PBS, then their polarisations are rotated by a $\lambda/2$ wave-plate to give a roughly equal mixture of $\sigma_\pm$ and π in both beams. The two 397 nm beams ($397D40$

Table 5.4 Laser beam spot sizes and polarisations. The spot sizes are all $1/e^2$ radii. The 'saturation power' is the beam power that gives one saturation intensity

Beam	Spot size (μm)	Saturation power (μW)	Polarisation
397/40	42	2.6	10% $\sigma^\pm$, 90% π
397/43	42	2.6	90% $\sigma^\pm$, 10% π
866	105	1.6	85% $\sigma^\pm$, 15% π
854	82	1.0	15% $\sigma^\pm$, 85% π
850	80	1.0	σ^+ or σ^-
393	140	30	σ^+
397σ	70	7.2	σ^+
397π	70	7.2	π

and 397D43) are also combined on a PBS and rotated by a $\lambda/2$. This wave-plate angle is empirically optimised for the best ^{43}Ca$^+$ fluorescence, which occurs for 10% π, 90% $\sigma_\pm$. The spot sizes ($1/e^2$ intensity radius) are $w = 42\,\mu$m for the 397 nm beams, and $w \approx 100\,\mu$m for the 866 and 854 nm beams.

The three σ beams are at small angles to one other. The magnetic field is aligned along the 397σ path, so the 850 and 393 nm beams can never be purely σ polarised. However for a circularly polarised beam misaligned by $2.5°$ the π polarisation impurity is only 10^{-3} in intensity, which is unimportant. The $\lambda/4$ wave-plates are all on tip-tilt mounts adjusted to optimised the polarisation quality as measured by optical pumping. The 397 and 850 σ also have Glan–Taylor polarisers[24] before the wave-plates to clean the polarisation further.

5.5   Raman Lasers

Our Raman laser system needs to deliver several pairs of beams to the trap. The beam-pairs need to have a stable difference-phase over one shot of the experiment, and well-controlled intensities. The Raman beam difference frequencies need to be $\sim$5 MHz for driving transitions in ^{40}Ca$^+$ qubits or for driving light-shift gates in either isotope, or $\sim$3.2 GHz for driving transitions in ^{43}Ca$^+$ qubits.

---

[24]Thorlabs GT10-A and GT10-B for the UV and IR respectively.

5.5.1 Laser Sources

The laser sources are a pair of frequency-doubled amplified diode lasers.[25] A sample of the master system's fundamental output at $\sim$794 nm goes through a double-pass $f_{AO} \approx 800$ MHz AOM[26] (imparting a $\approx$1.6 GHz shift) and injects the slave system's diode. The slave diode injection-locks to the master laser such that the output of the master and slave diode are coherent. Amplifying and frequency-doubling the light out of the two diodes produces two coherent beams at $\sim$397 nm with a frequency difference of $4f_{AO} \approx 3.2$ GHz. We follow this approach rather than directly producing a shift of 3.2 GHz in the blue, because there are no efficient modulators available for such high frequencies. With this arrangement we are relying on interferometric stability of the injection path and both doubling cavities: it is not obvious that the resulting differential phase noise will be tolerable. We investigate this in Sect. 6.4 and find that the phase noise is small enough not to be a serious issue for most experiments.

The absolute frequency of the Raman beams is not that important, so we do not lock it. We find that the free-running master laser drifts at the 100 MHz level; this is negligible compared to our typical Raman detunings of 500 GHz–4 THz. We can adjust the frequency output of the lasers between detunings of 0 to -4 THz from the $S_{1/2}$–$P_{1/2}$ transition whilst only requiring moderate adjustments to the doubling cavities (taking $\lesssim 10$ min) to keep the output power above $\sim$100 mW for the master and $\sim$60 mW for the slave. If we optimise the doubling cavities for one particular frequency we can achieve output powers of 130 mW from the master and 80 mW from the slave. Further details of the injection-locking of these systems have been published separately [LBL13, Lin12].

5.5.2 AOM Switch-Yard and Optical Fibres

For the experiments in this thesis we use up to three Raman beams. As the master has a higher-power output, we supply two of the beams from the master and one from the slave. Typically one of the beams needs to be scanned in frequency (e.g. to address motional side-bands), and the other two operate at a fixed frequency.

The fixed-frequency beams are switched by 200 MHz single-pass AOMs, and the scanned beam by a double-pass 85 MHz AOM.[27] The single-pass AOMs give high diffraction efficiency ($\sim$85%), but due to the diffraction angle changing with frequency the beam frequency can only be scanned by $\sim \pm 3$ MHz before the fibre input-coupling is misaligned (see Fig. 5.5) – this reduces the coupled power, but more importantly increases the sensitivity to beam pointing noise originating from the doubling cavities. The double-pass AOM has a lower efficiency ($\sim$60%), partially

---

[25]Toptica TA/DL-SHG110 Pro (Master)/Normal (Slave).

[26]Brimrose TEF-800-300-794.

[27]IntraAction ASM-2001.5B8 (200 MHz) and ASM-851.5B8 (85 MHz).

Fig. 5.5 Optical power after the fibre versus AOM frequency shift for a single-pass 200 MHz AOM

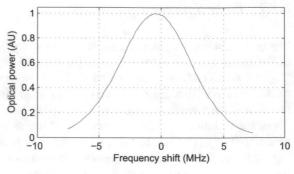

Fig. 5.6 Optical power after the fibre versus AOM frequency shift for a double-pass 85 MHz AOM

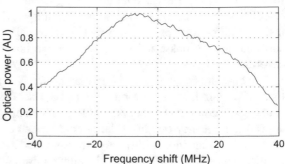

due to the complicated beam geometry needed,[28] but allows tuning over ± 40 MHz of frequency shift (see Fig. 5.6).

To decouple beam pointing changes from doubling cavity adjustments, pointing noise from the doubling cavity lock piezo, and AOM frequency shifts, we use short (1m) optical fibres. A potential issue is optical damage to the fibre tips from the ~ 30 mW UV beam focussed to $\sim 2\,\mu$m (the fibre mode-field diameter). We avoid this problem by having the fibres 'endcapped' by the manufacturer[29] – this involves fusing $200\,\mu$m of unstructured glass onto the fibre tip, so that the intensity at the air-glass interface is substantially reduced. We have operated these fibres with up to 100 mW at the input face (70% transmission) with no noticeable optical damage or reduction in transmission over many months.

The RF chain for each AOM consists of a DDS source, a DC biased mixer,[30] a 250 MHz low-pass filter,[31] a switch, and a power amplifier.[32] Fast pulse-shaping is implemented by the DDS source. The DC biased mixer is used for slow amplitude modulation and is driven by the noise-eater. The beam on / off ratio (measured after

---

[28]The AOMs only have a reasonable diffraction efficiency for vertically polarised light, so a polarisation quadrature scheme cannot be used.

[29]Schäfter+Kirchoff PMC-E-360Si-2.3-NA012-3-APC.EC-150-P.

[30]ZP-3MH+ : bias current of 0–10 mA injected into the IF port.

[31]SLP-250+.

[32]ZHL-03-5WF.

the fibre) is very good – for 10 mW out of the fibre with the AOM on, we get < 0.1nW with the AOM off: an on / off ratio of $< 10^{-8}$ (with the measurement limited by the power meter resolution).

5.5.3 Beam-Paths

As we wish to maximise the Raman beam intensity we want a small beam waist at the ion. The main limitation on decreasing the beam waist is beam-pointing drift and noise. To reduce these issues we minimise the length of beam-path and number of optics between the fibre output and the ion, and mount all the optics as rigidly as possible. The fibre output couplers are constructed[33] to give collimated beam diameters of $d = 1.05$ mm ($f = 7.5$ mm) or $d = 2.18$ mm ($f = 15$ mm).

Following polarisation clean-up with a PBS, and power measurement with a pick-off and photodiode, there are two mirrors and a wave-plate before the $f = 200$mm focussing lens, giving a total optical path length from fibre output to ion of $\sim$50cm. The beam waist at the ion is $w = 60\,\mu$m ($f = 7.5$mm) or $w = 27.5\,\mu$m ($f = 15$mm). The final mirror before the lens has a piezo-actuated mount[34] driven by an open-loop piezo controller.[35] This enables us to align the beam onto the ion much more precisely than by hand, greatly reducing the sensitivity to subsequent pointing drift and noise (see Sect. 6.4.1).

The Raman beam directions, polarisations and notation used in this thesis are given in Fig. 5.7. Which beams are driven from which laser and AOM differ depending on experiment – this is easy to change as it just involves swapping fibres.

The normal arrangement for working on the ^{43}Ca$^+$ stretch qubit is for R_V and $R_\parallel$ to be driven by the master laser, and for R_H to be driven from the slave, with $R_\parallel$ driven by the double-pass AOM. This allows for motionally-insensitive (R_V, R_H) and motionally-sensitive ($R_\parallel, R_H$) driving of the qubit, and implementation of the light-shift two-qubit gate ($R_V, R_\parallel$).

The arrangement for motionally-insensitive driving of the ^{43}Ca$^+$ clock qubit, used for the single-qubit benchmarking of Chap. 7, is for $R_\parallel$ to be driven by the master, and for $R_{\parallel 2}$ to be driven by the slave.

---

[33]Fibre coupler construction (Thorlabs part numbers): FC/APC fibre chuck (SM05FCA) in 0.5" lens-tube (SM05M10) with a $f = 7.5$mm (AC050-008-A-ML in S05TM09 adapter) or $f = 15$mm (AC064-015-A-ML) focal length collimating lens, mounted in a lens-tube clamp (SM05TC).

[34]Initially home-made mounts, later work with 'Radiant Dyes MDI-H-2-1" with Piezo-drive'.

[35]Thorlabs MDT693B.

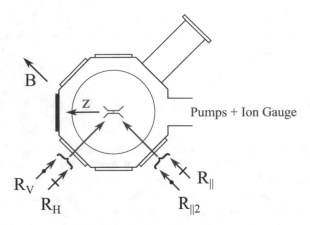

Fig. 5.7 Geometry of the Raman beams with respect to the quantisation axis B, including polarisations. The two beams incident on each window are co-axial

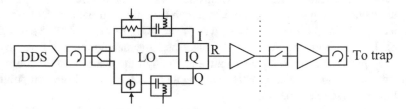

Fig. 5.8 Schematic of the microwave qubit drive. Parts to the *left* of the *dashed line* comprise the SSB source. The DDS source at $\sim$300 MHz is mixed with a 3.5 GHz LO. The phase and amplitude shifters are adjusted to null the *upper* sideband. The mixer bias currents are chosen to null the carrier leakage

5.6 Microwave Qubit Drive

Transitions in the ground level of ^{43}Ca$^+$ are driven by a DDS ($f_{\text{clk}} = 1$GHz) mixed up to 3.2 GHz with a single-sideband system.

The single-sideband source (Fig. 5.8) consists of an isolated[36] DDS source[37] operating at $f_{\text{IF}} \sim 280$ MHz driving an IQ mixer network. The LO port of the mixer is driven at $f_{\text{LO}} = 3.5$ GHz, $P_{\text{LO}} = 10$ dBm. The voltage variable attenuator[38] and voltage variable phase-shifter[39] are adjusted to null the upper sideband, giving a principal output at $f_{\text{RF}} = f_{\text{LO}} - f_{\text{IF}}$. A pair of bias-tees[40] are used to inject small currents (± 65 mV via a 50Ω resistor) into the I and Q ports to balance the mixer,

---

[36]Ocean Microwave 165846-101.

[37]Analog Devices AD9910 evaluation board.

[38]Minicircuits ZX73-2500M-S+.

[39]Minicircuits JSPHS-42.

[40]Minicircuits ZFBT-4R2G+.

and hence reduce LO leakage. A low-noise amplifier[41] increases the output level to ≈ -2 dBm. After optimising the nulling, the amplitudes of the suppressed sidebands relative to the desired -1 sideband are $\{-2, 0, +1\} = \{-44, -55, -55\}$dBc. Changing the DDS frequency (and hence the output frequency) by 5 MHz changes the sideband amplitudes to $\{-2, 0, +1\} = \{-44, -53, -24\}$ dBc.

The output from the SSB source, switched by the experimental control computer, drives a power amplifier[42] connected to the trap via an isolator. At full output power 41.3 dBm (13.4 W) is fed to the trap.

The microwaves are fed in to the vacuum system via a 'DC' feed-through connected to one of the grounded trap blades, the other end of the blade is connected to trap ground. The intent of this is to get a microwave current flowing near the ions in order to maximise the oscillating magnetic field that drives the qubit transitions. The return loss of the microwave input is 1.1 dB: 80% of the incident power is reflected. It is unclear how much is directly reflected by the feed-through, and how much microwave current flows through the trap blade.

We measure a microwave polarisation of $\{\sigma^+, \pi, \sigma^-\} = \{0.25, 0.63, 0.73\}$. At full power, 13.4 W at the feed-through, the $S_{1/2}^{4,+4}$–$S_{1/2}^{3,+3}$ (σ^-) stretch qubit Rabi frequency is 83 kHz. The microwave drive extinction is 74 dB in power, as measured from the clock transition Rabi frequency with the microwave switch on and off.

5.7 Experimental Control

The experiment is controlled by several different computers. The main experimental control system consists of a real-time DOS system interfaced, via a serial link, to a desktop computer running Linux. The Linux system handles (slow) communication with the main parts of the experiment; communicating over GPIB, Ethernet, and USB to DDS systems, RF synthesizers, DACs, etc. The real-time DOS system runs the pulse sequence (100 ns timing resolution, 6 μs 'dead time' between pulses) and counts clicks from the PMT. In many experiments a higher timing resolution is needed for the coherent operations. For these the DOS system pulse-sequencer is used for state-preparation and read-out pulses, and triggers an FPGA based pulse-sequencer which implements the coherent operations.

5.7.1 Coherent DDS

The 'Coherent DDS' is a 4 channel phase coherent DDS system with pulse-shaping. This is the RF source that provides the RF for the Raman laser AOMs. Each channel has 8 frequency/phase/amplitude profiles, and a programmed pulse-shape. TTL sig-

---

[41]Minicircuits ZX60-362GLN-S+.

[42]Minicircuits ZHL-16W-43S+.

nals from the pulse-sequencer select which profile is active, and trigger the playback of a rising or falling pulse-shape. The DDS is coherent in that the 4 channels are phase-locked – two channels with the same programmed frequency and phase will be edge aligned – and that the phase reference is not lost on switching a channel to a different frequency temporarily – if a channel is switched from one frequency to another and back again, the phase offset is the same as if the channel did not switch. Further technical details on the DDS can be found in Appendix B.1.

5.7.2   Laser Intensity Stabilisation

The laser intensity 'noise-eater' works by keeping constant the beam power hitting a photo-diode by adjusting the RF drive amplitude of the AOM earlier in the beam path. The laser intensity stabilisation is implemented by an FPGA card[43] in a computer separate from the rest of the experimental control system. The PID feedback loop is implemented in the FPGA, while power set-point adjustment and loop saturation monitoring are performed by the computer.

The FPGA also receives a TTL signal from the experimental control computer that opens the feedback loop while holding the AOM drive amplitude. For the Doppler cooling and optical pumping beams the feedback loop runs at 100 kHz and the noise-eating is enabled by the same signal that switches the AOM on and off. For the Raman beams the feedback loop runs at 500 kHz and the noise-eating is enabled by an additional signal from the experimental control computer. The Raman beams and Raman noise-eater are turned on for 300 µs at the start of each pulse sequence (before state preparation) to stabilise the beam power, but the powers are not actively stabilised during the Raman manipulations themselves. This is because the Raman manipulations frequently consist of short pulses ($\sim$µs). Turn-on transients in the photodiodes and signal processing circuitry mean that the signal does not stabilise until about 2 µs after switching the beam on, hence we cannot stabilise the intensity of pulses shorter than $\sim$5 µs. Rather than risk systematic effects that depend sensitively on the pulse length profile of each sequence, we prefer to tolerate the potential slow thermal transients over the course of the sequence. As we measure in Chap. 7, these are not a significant problem.

References

[Gul03]   Gulde, S. 2003. *Experimental Realization of Quantum Gates and the Deutsch-Jozsa Algorithm with Trapped Ca40 ions*. PhD thesis, University of Innsbruck.
[All11]   Allcock, D.T.C. 2011. *Surface-Electrode Ion Traps for Scalable Quantum Computing*. PhD thesis, University of Oxford.

---

[43]National Instruments PCIe-7852R.

[SSWH12] Siverns, J.D., L.R. Simkins, S. Weidt, and W.K. Hensinger. 2012. On the application of radio frequency voltages to ion traps via helical resonators. *Applied Physics B* 107 (4): 921–934.

[ZB61] Zverev, A.I., and H.J. Blinchikoff. 1961. Realization of a filter with helical components. *IRE Transactions on Component Parts* 8 (3): 99–110.

[LAS+12] Linke, N.M., D.T.C. Allcock, D.J. Szwer, C.J. Ballance, T.P. Harty, H.A. Janacek, D.N. Stacey, A.M. Steane, and D.M. Lucas. 2012. Background-free detection of trapped ions. *Applied Physics B*.

[Szw09] Szwer, D.J. 2009. *High Fidelity Readout and Protection of a 43Ca+ Trapped Ion Qubit*. PhD thesis, University of Oxford.

[Lin12] Linke, N.M. 2012. *Background-free detection and mixed-species crystals in micro- and macroscopic ion-traps for scalable QIP*. PhD thesis, University of Oxford.

[LBL13] Linke, N.M., C.J. Ballance, and D.M. Lucas. 2013. Injection locking of two frequency-doubled lasers with 3.2 GHz offset for driving Raman transitions with low photon scattering in 43Ca+. *Optics letters* 38 (23): 5087–9.

Chapter 6
Experiment Characterization

It is important to understand the behaviour and limitations of the apparatus. This chapter describes the characterisation of some of the more important aspects of the experimental apparatus.

6.1 Trap Behaviour

6.1.1 Axial Micro-Motion

As we saw in Sect. 5.1 we expect some axial RF pseudo-potential as we are driving the blades asymmetrically. This axial pseudo-potential causes two interrelated issues. The first is that the axial RF pseudo-potential adds to the axial DC potential, hence if the RF amplitude changes the axial trap frequency will change. This is a problem as we require the axial trap frequency to be very stable (~ 10 Hz) for our experiments. It is easy to make the DC voltages that generate most of the axial trap frequency stable, but it is far harder to make the RF amplitude stable.

The second effect is that ions displaced from the centre of the trap along the trap axis will have driven motion (micro-motion). Thus the ions in the axial crystals we use will each have different micro-motion amplitudes. The primary effect of this is a reduction in Rabi frequency for any beam that 'sees' the axial motion (i.e. a beam that has a projection along the trap axis).

To confirm our trap model we measure the change in the axial trap frequency as a function of the radial trap frequency. From Eq. 2.6 and the coefficients in Table 5.1 we find

© Springer International Publishing AG 2017
C.J. Ballance, *High-Fidelity Quantum Logic in Ca$^+$*,
Springer Theses, DOI 10.1007/978-3-319-68216-7_6

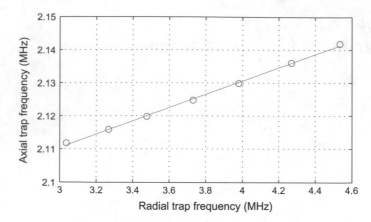

Fig. 6.1 Measured dependence of axial mode frequency on radial mode frequency for a single ^{40}Ca$^+$ ion. The linear fit gives a cross coupling coefficient $\frac{\partial \omega_z}{\partial \omega_\perp} = 20.2$ kHz/MHz

$$\frac{\partial \omega_z}{\partial \omega_\perp} = \frac{\omega_\perp}{\omega_z} \left(\frac{\alpha_z}{2z_0^2} \right) \left(\frac{\alpha_r}{\rho_0^2} \right)^{-1}$$
$$= 9.7 \, \text{kHz/MHz} \qquad (6.1)$$

where the equation is evaluated for $f_z = 2.13$ MHz and $f_\perp = 3.8$ MHz. We measure this relationship by finding the (higher frequency) radial mode and axial mode frequencies at several different trap RF amplitudes (Fig. 6.1). The measured cross-coupling coefficient of 20.2 kHz/MHz differs significantly from the predicted value of 9.7 kHz/MHz.

We can also measure the axial pseudo-potential by measuring the axial micro-motion amplitude of a single ion as we displace it along the axis of the trap. The amplitude of this axial micro-motion is

$$\boldsymbol{u} = \frac{1}{2} q_z z \hat{\boldsymbol{z}} \qquad (6.2)$$

where q_z is the Mathieu 'little q' parameter (Sect. 2.2). This driven motion gives rise to sidebands at $\omega_0 + n\Omega_{\text{RF}}$, where ω_0 is the qubit frequency [BMB+98]. The sideband amplitude is $J_n(\eta_{\text{eff}})$, where $\eta_{\text{eff}} := \Delta \boldsymbol{k} \cdot \boldsymbol{u}$. We measure the ratio of the sideband to carrier Rabi frequency using a pair of Raman beams with $\Delta \boldsymbol{k} \propto \hat{\boldsymbol{z}}$ for a range of axial positions of the ion (Fig. 6.2). We displace the ion by applying a differential voltage to the end-caps. We calibrate the displacement by fitting images of the ion, having previously calibrated the camera magnification to be 1.244 px/μm by fitting an image of a two-ion crystal with a known axial frequency (and hence a known ion-ion separation). From this data we calculate the micro-motion amplitude as a function of displacement, and find that $|\Omega_{\text{SB}}/\Omega_0|/z = 0.24 \, \mu\text{m}^{-1}$ for $f_\perp = \{3.74, 4.24\}$ MHz and $f_z = 1.93$ MHz. For a two-ion crystal with axial centre-of-mass frequency

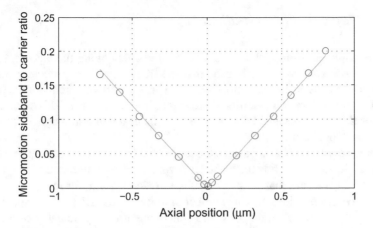

Fig. 6.2 Micro-motion sideband to carrier Rabi frequency versus axial displacement. The linear fit gives $|\Omega_{SB}/\Omega_0|/z = 0.24\,\mu m^{-1}$

2 MHz centred in the trap each ion has a micro-motion amplitude of 38 nm, leading to a carrier Rabi frequency of $0.83\Omega_0$ and a micro-motion sideband amplitude of $0.38\Omega_0$. From the measured trap frequencies and the simulated trap geometry we calculate $\eta_{\text{eff}}/z = 0.34\,\mu m^{-1}$, a poor agreement.

So, both of our measurements of the axial micro-motion are in disagreement with our predictions – do they agree with each other? From our axial-radial cross-coupling measurement we find $q_z/q_r = 0.106$ (compared to the expected value of 0.072). Thus for the parameters in our micro-motion gradient measurement we expect $q_z = 0.045$ and hence $|\Omega_{SB}/\Omega_0|/z = 0.25\,\mu m^{-1}$, in good agreement with the measured value $0.24\,\mu m^{-1}$. Thus our trap has a different q_z/q_r than expected from our model. As q_z/q_r only depends on the trap geometry parameters this means our trap is not built as designed. From separate measurements we know $\alpha_z = 0.292$ (modelled value 0.285), hence it seems $\alpha_r = 0.344$ (modelled value 0.495). The reduction in RF blade 'efficiency' could be caused by the blades being narrower or further away from the ion than designed. When the trap was assembled the blades were polished with fine emery cloth to improve the finish – this could cause both a narrowing of the blade tip and a larger ion-blade separation. Measurements on a spare blade that was also polished suggest that the ion-blade separation could easily have been increased by the $\sim 100\,\mu m$ necessary to achieve the measured α_r. If this assessment is correct, we are applying an RF amplitude of 1.2kV zero-peak to the blades for our typical radial trap frequencies, rather than the expected 830V.

6.1.2   Magnetic Field Gradient

We have observed a substantial magnetic field gradient along the axis of the trap. This is a significant issue as it means that the frequency of magnetically sensitive transitions depends on the position of the ion in the trap. This gradient is large enough to cause significant frequency shifts between ions in a small crystal. This field gradient is unexpected, as we designed the ion trap to be made out of non-magnetic materials to avoid such problems.

After we observed what we suspected to be a field gradient, we sought to measure it precisely. To do this we rely on the fact that we can measure the axial trap frequency accurately. This means that we can calculate the ion spacing for many-ion crystals precisely. We can then load varying number of ions into an axially weak trap, measure the resonant frequencies of the ions, and hence measure the gradient accurately.

In our experiments we use the 'stretch' qubit of $^{43}Ca^+$. We set $f_z = 475.9(1)$ kHz and scan the frequency of the microwave probe field ($t_\pi = 160\,\mu s$) over the range of splittings of the qubits. Assuming the lab magnetic field stays constant over the frequency scan we can fit the experimental data for the difference in qubit frequency between the different ions. An example scan is plotted in Fig. 6.3. We repeat this experiment for 2, 3, and 4 ions at our usual magnetic field of 2.00 G, and for 4 ions at a field of 3.60 G. The results (Fig. 6.4) are consistent with a linear gradient in qubit frequency of $1.404(1)$kHz/μm at our usual field, and a slightly reduced gradient of $1.375(5)$ kHz/μm at the higher field. This qubit frequency gradient corresponds to a magnetic field gradient of $\frac{dB}{dz} = 0.573(1)$ mG/μm. For a two-ion crystal with $f_z = 2$ MHz the ion-spacing is $3.5\,\mu$m, corresponding to a 'stretch' qubit frequency difference of 4.9 kHz.

To check for gross variation in the qubit frequency over the length of the trap we move the ion $\pm 50\,\mu$m along the trap axis (Fig. 6.5). There is no remarkable deviation from a linear gradient. The slight curvature is potentially from a barrel distortion in the imaging system.

In other experiments we find that the qubit frequency gradient does not change with radial trap frequency. This means that the gradient we see cannot be caused by spatially varying AC Stark shifts from currents flowing at the trap RF frequency, and hence must caused by a static magnetic field gradient. As the gradient is not dependent on the applied magnetic field this shows it is not a 'soft' magnetic effect. We believe that this gradient must be caused by some magnetized material in the trap. As there is no large offset field (the field at the ion's position is $\sim$1 G with no applied magnetic field) the magnetic material must be close to the ion. This suggests that the blades may be the source. The blades were milled out of steel. We intended the blades to be made out of non-magnetic 316 L grade steel, but cannot confirm that they were made out of this grade. Furthermore, the blades were not annealed after machining. To test the hypothesis that the blades are the source of the gradient, we measured the field from a spare blade machined at the same time and out of the same stock as the blades used in the trap. Contacting the blade to a Hall effect sensor

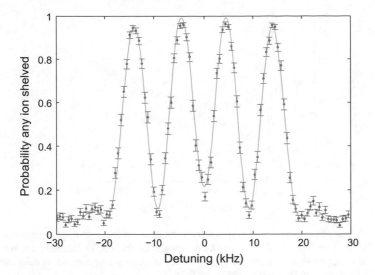

Fig. 6.3 Measurement of the qubit frequency gradient. A 4 ion crystal is probed with a frequency-scanned microwave pulse. Due to the qubit frequency gradient, each of the ions comes into resonance at a different frequency. The outer ions in the crystal are approximately 10 μm from the trap centre

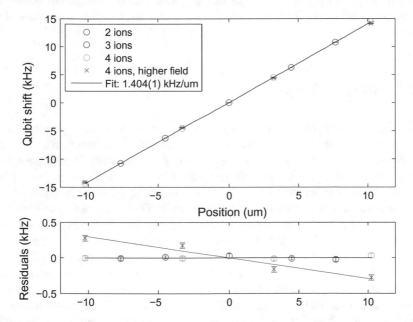

Fig. 6.4 Precise measurement of the ^{43}Ca$^+$ 'stretch' qubit frequency gradient. The frequency error bars are 1σ fit uncertainties. The ion position error bars are negligible. The measurements taken at the normal field of 2.00 G (*circles*) are consistent with a linear gradient (see residuals). The measurements taken at a higher field (3.60 G, crosses) are consistent with a slightly reduced gradient (smaller by $-0.029(5)$ kHz/μm)

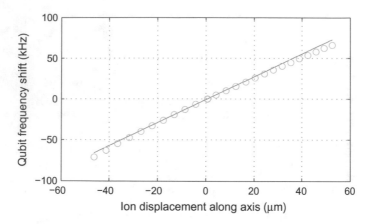

Fig. 6.5 Coarse measurement of the qubit frequency variation for large displacements from trap centre. The axial position is calibrated by fitting images of the ion. The absolute scale is set by fitting an image of two ions in a trap with known axial frequency. The *blue line* is calculated from the gradient measured at centre of the trap (Fig. 6.4)

gave an indicated field of ~3 G. This is a very crude measurement, but shows that the magnetic field from the blades is of the right scale to cause the axial gradient we observe.

6.1.3 Recrystallisation

When there is a background gas collision or one of the Doppler cooling lasers glitches the ions can be heated far above the Doppler limit. After a disturbance a single ion will be reliably cooled back down to the Doppler limit. However, if there are several ions in the trap the dynamics is much more complicated. Once the ions heat up and decrystallise they experience a Doppler shift larger than the small Doppler cooling laser detuning, and hence are not cooled effectively. The ions typically reach a 'hot' stable state – this can be detected by a drop in fluorescence. In our experiment, for a two-ion ^{43}Ca$^+$ crystal, decrystallisations happen ~4 times an hour on average.

The experimental control computer monitors the ion's fluorescence in a 'check' bin at the end of each shot of the experiment. If the number of photon counts is less than the 'all ions bright' threshold the computer pauses the sequence and repeats the measurement up to 100 times, or until the threshold is met. If the threshold has not been met after 100 repeats the recrystallisation routine runs. This avoids triggering the recrystallisation routine if there was only a transient event, such as a minor laser frequency excursion.

To recrystallise the ions we reduce the axial and radial trap strengths to $f_z = 500$ kHz and $f_r = 1$ MHz, and red-detune the Doppler cooling beam by 150 MHz.

We then wait 500 ms, go back to the original trap frequencies, and scan the Doppler cooling laser back to its original detuning. This procedure reliably recrystallises small crystals of $^{40}Ca^+$, $^{43}Ca^+$, and mixed-species crystals (only cooling one species).

6.2 State-Preparation and Readout

In this section we discuss the typical parameters used and results achieved for state-preparation and readout in the different $^{43}Ca^+$ qubits we use, as well as for the $^{40}Ca^+$ qubit.

6.2.1 $^{43}Ca^+$ Optical Qubit

To readout any of the $^{43}Ca^+$ qubits we first need to be able to read out the 'optical' qubit, that is, discriminate between population in the ground level ($4S_{1/2}$) and shelf level ($3D_{5/2}$). To do this well we need to determine the ion's state fast compared to the $D_{5/2}$ decay time-constant of $\sim$1.2 s. For this we need a high fluorescence count-rate from the ion, and a low background count-rate from laser scatter.

With the Doppler cooling parameters optimised for high fluorescence the ion signal count-rate is 52.4 ks^{-1}, and the background count-rate 1.4 ks^{-1}. This fluorescence signal corresponds to a $P_{1/2}$ population of 13.5%, and is achieved by using the 397/40 laser to Doppler cool in addition to 397/43 laser. The 397/43 EOM sideband-to-carrier ratio is 0.7 at a drive frequency of 3.243 GHz, with a total beam power of 80 μW ($30I_0$). The 397/43 carrier frequency is tuned -1.38 GHz from the $^{40}Ca^+$ resonance, which is slightly red-detuned from the $S_{1/2}^3 \leftrightarrow P_{1/2}^4$ transition. The 397/40 is tuned to $+1.78$ GHz from the $^{40}Ca^+$ resonance with a beam power of 15 μW ($5I_0$) – this is near the $S_{1/2}^4 \leftrightarrow P_{1/2}^4$ transition which the 397/43 EOM sideband also addresses. The 866 repumper laser is tuned -3.30 GHz from the $^{40}Ca^+$ resonance with a beam power of 1.6 mW ($10^3 I_0$).

Using the near-optimum photon-counting bin length of 500 μs the optical qubit readout errors are $\epsilon_S = 0.4(2) \times 10^{-4}$ and $\epsilon_D = 2.9(6) \times 10^{-4}$, giving an average readout error of $\bar{\epsilon} := \frac{1}{2}(\epsilon_D + \epsilon_S) = 1.7(3) \times 10^{-4}$. We note that, if desired, we could reduce this error to $\lesssim 10^{-4}$ by using time-resolved detection [MSW+08].

6.2.2 $^{43}Ca^+$ Stretch Qubit

To read out the stretch qubit we need, amongst other beams, some 850 nm repumpers. Rather than using separate σ and π beams we use a single circularly polarised beam slightly misaligned with the quantisation axis to give a weak π component. This

trick slightly lowers the achievable readout fidelity – this is carefully analysed in
[Szw09]. We use a 393 nm shelving time-constant of 33 μs and a shelving pulse
length of 400 μs (during which both the 850 and 393 beam are applied) – these are
near the optimum. The 850 beam power is 1 mW ($10^3 I_0$).

The combined state-preparation and readout error, including the optical qubit
readout error, is $\epsilon_\uparrow = 6.9(7) \times 10^{-4}$, $\epsilon_\downarrow = 11.1(8) \times 10^{-4}$, giving $\bar{\epsilon} = 9.0(5) \times 10^{-4}$.
The theoretical minimum shelving error is $\bar{\epsilon} = 4 \times 10^{-4}$, giving a theoretical total
error from shelving and optical qubit readout of $\bar{\epsilon} = 5.7 \times 10^{-4}$. From these results
we can put a limit on the *manifold* preparation error of $\bar{\epsilon} \lesssim 3 \times 10^{-4}$, but cannot
say anything firm about the state preparation error as the 'state-selective' shelving
reliably shelves $S_{1/2}^{4,+3}$ (the state that imperfect state preparation is most likely to
populate) as well as the desired state $S_{1/2}^{4,+4}$.

6.2.3 $^{43}Ca^+$ Low-Field Clock Qubit

We can prepare the clock qubit in two ways. We can optically pump to the stretch
state, then use 4 microwave or Raman π-pulses to map to the lower state of the
clock qubit, or we can directly optically pump to the lower state of the clock qubit
using 397 nm π light (using the selection rule $M'_f = 0 \nrightarrow M_f = 0$ if $\Delta F = 0$).
The direct optical pumping seems the more elegant solution, but has two problems.
The first is an off-resonant allowed transition, $S_{1/2}^{4,0} \leftrightarrow P_{1/2}^{3,0}$, that limits the state
preparation fidelity to 98.9% [Szw09]. The second is due to the inefficiency of the
state preparation – we have to scatter many photons to get to the target state, even
if we start from a state very close to the target state. The recoil from these scattered
photons heats the ion's motion. This means that sideband cooling is slow, and the
final temperature achieved is relatively high. The problem with preparing the clock
qubit with microwave π pulses is the limited Rabi frequency available combined
with the qubit frequency gradient: for 2 ions the total preparation error is ~1%. We
could use Raman π pulses to perform the mapping, but this would require another
Raman beam path, and lower the power available in the other Raman beams.

To read out the clock qubit we have two options. The first is mapping one of
the qubit states back to the stretch state, while leaving the other in $S_{1/2}^3$, then using
the high-quality readout of the stretch state. The second is applying the manifold-
selective shelving directly to the population in the two qubit states. This has a much
higher theoretical minimum error ($\bar{\epsilon} = 0.4\%$) than the stretch state shelving as we
no longer have an approximate cycling transition via $P_{3/2}^{5,5}$ [Szw09]. We choose the
second method for experimental simplicity.

For state preparation by optical pumping, we adjust the beam power to minimise
the error with a 2 ms pumping pulse. The combined state-preparation and readout
error is $\bar{\epsilon} = 1.8\%$. If we start in the stretch state at $\bar{n} = 0$, the final temperature

after the optical pumping is measured to be $\bar{n} = 0.3$. Using microwave π pulses after stretch state preparation the combined state-preparation and readout error is $\bar{\epsilon} = 1.3\%$ for a single ion.

6.2.4 $^{40}Ca^+$ Zeeman Qubit

For $^{40}Ca^+$ readout we use a 393 shelving time-constant of $\sim 7\,\mu s$, an 850 power of $12\,mW$ ($1.2 \times 10^4 I_0$), and a shelving pulse length of $30\,\mu s$. For a single ion the combined state-preparation and readout error is $\epsilon_\uparrow = 7.2\%$ and $\epsilon_\downarrow = 2.6\%$, giving $\bar{\epsilon} = 4.9\%$, close to the theoretical limit of $\bar{\epsilon} = 3.6\%$ [McD03]. This readout process involves placing a 'dark resonance' on the shelving transition for one of the qubits. By using a large 850 intensity this feature has a reasonable width, however micro-motion sidebands on the readout lasers still decrease the effectiveness of the dark resonance, increasing the readout error. For two ions the axial micro-motion increases the combined state-preparation and readout error to $\bar{\epsilon} = 7.5\%$.

6.3 Qubit Dephasing and Frequency Drift

All the qubits we use are magnetically field sensitive to some degree. As we do not use any magnetic shielding, any modulation of the laboratory magnetic field is seen directly by the ion. The uncontrolled laboratory magnetic field has noise (frequency components higher than the sequence repetition rate) that decoheres the qubit, and drift that causes time varying modulation of the qubit frequency.

6.3.1 Magnetic Field Drift and Modulation

A significant source of magnetic field modulation is the 50 Hz mains power network. We characterise this by measuring the variation of the stretch qubit frequency as a function of delay from the mains zero-crossing. We trigger the sequence on the mains zero crossing, delay for a while, then perform a $50\,\mu s$ Ramsey experiment to determine the qubit frequency (Fig. 6.6). We observe peak-to-peak variations of 5 kHz in qubit frequency, with significant structure from the harmonics of 50 Hz. For careful experiments we adjust the length of the sequence so that the critical parts of the sequence start $\sim 6\,ms$ after the line-trigger, as around this time the qubit frequency varies slowly with time. If this 50 Hz noise caused a significant problem it could be removed by a line-triggered feed-forward circuit that applied a modulation to the magnetic field servo coils to cancel out the laboratory magnetic field.

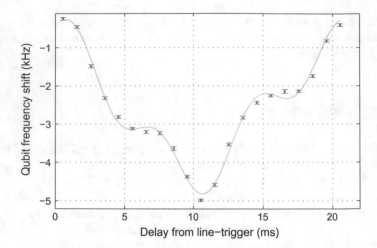

Fig. 6.6 Stretch qubit frequency variations with (50 Hz) mains phase. The data are the measured qubit frequency at different delays from the mains zero-crossing. The solid line is a fit of 3 sinusoids with frequencies {50, 100, 150} Hz. The fitted amplitudes are {1.7, 0.1, 0.65}kHz = {0.7, 0.04, 0.27} mG

The long-term laboratory magnetic field (and field seen by the ion) is inherently very stable – in the early morning when no-one[1] is in the building the magnetic field variations are below 0.1 mG. In addition, we see no drift in the magnetic field from heating the oven up to load, as we have seen in a previous trap [Hom06]. However, during the day a nearby goods lift and a selection of scanned superconducting magnets change the laboratory magnetic field by up to 3 and 15 mG respectively with slew-rates of up to 3 mG per minute. These slew rates are large enough that even after running the magnetic field servo before the start of each scan significant errors ($\sim 10^{-3}$) can be caused by the magnetic field variation.

6.3.2   Stretch Qubit Coherence

We measure the coherence time of the stretch qubit, using a series of microwave Ramsey experiments, to be $\tau = 3.2(5)$ ms (Fig. 6.7) when the sequence is line-triggered. Using a single spin-echo increases the coherence time considerably, showing that the the magnetic field noise is strongly correlated over the sequence length. For small errors the spin-echo contrast decay is quadratic, $\epsilon = (t/\tau)^2$, with $\tau = 9$ ms.

---

[1] Apart from a magnetically field sensitive D.Phil. student.

Fig. 6.7 Stretch state
coherence time measurement
when line-triggering the
sequence, probing with
microwaves. The exponential
fit gives a coherence time of
$\tau = 3.2(5)$ ms

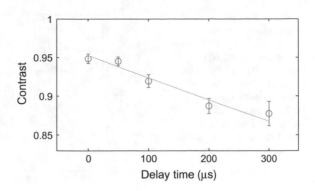

Fig. 6.8 Clock state
coherence time
measurement, probing with
microwaves. The exponential
fit gives a coherence time of
$\tau = 6(1)$ s

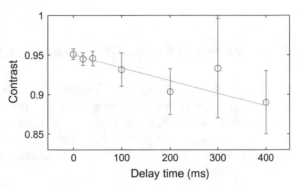

6.3.3   Clock Qubit Coherence

Due to the much reduced magnetic field sensitivity of the low-field clock qubit
(4.8 Hz/mG at $B_0 = 2$ G versus 2.45 kHz/mG for the stretch qubit) we expect a
much longer coherence time than the stretch state. Using a microwave probe field
we measure a coherence time of $\tau = 6(1)$ s (Fig. 6.8). As the microwave source
coherence time is very much longer than 1 s this decay is due to qubit decoherence.[2]

6.4   Raman Lasers

In this section we discuss Raman beam intensity and phase stability. We are concerned
about both short-term and long-term variations in the Raman beam intensity. Long-
term drifts of the intensity necessitate frequent recalibration of pulse areas. Short-term
variations (over the course of one shot of the experiment) lead to decoherence.

---

[2]We initially measured an anomalously small coherence time of ∼200 ms. The cause of this was a
faulty RF synthesiser with an intermittently locking PLL. We can thus claim to have built an atomic
clock with higher performance than a broken RF synthesiser from the early 1980s.

Fig. 6.9 Raman beam-pointing piezo hysteresis, using the Radiant Dyes piezo-actuated mirror mounts. We scan the piezo voltage in both directions to scan the beam ($w = 27\,\mu\text{m}$) over the ion, with a pulse area of $\pi/2$ when well aligned. There is a significant hysteresis, but it is repeatable

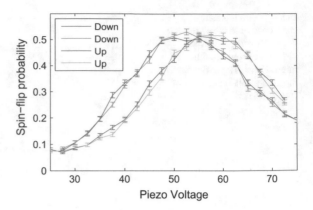

6.4.1 Beam-Pointing and Rabi Frequency Stability

To minimise the effect of small pointing fluctuations on the Raman beam intensity we need to accurately centre the beam on the ion. For the beam waist we use for our most critical experiments, $w_0 = 27\,\mu\text{m}$, a $1\,\mu\text{m}$ beam positioning drift in one of the Raman beams gives a 1.4×10^{-3} Rabi frequency fractional error if the beam was originally centred, or a 1.5×10^{-2} error if the beam was originally $5\,\mu\text{m}$ off (5% below the peak Rabi frequency).

We centre the beams on the ion using piezo-actuated mirror mounts. These allow us to scan the spot position at the ion over $\sim100\,\mu\text{m}$ (Fig. 6.9), albeit with significant hysteresis ($5\,\mu\text{m}$ at the ion). We align the beams by hand to within $\sim10\,\mu\text{m}$, then optimise the beam pointing using the piezos. We scan the piezos over a small range monitoring the Rabi frequency with a Raman carrier pulse of area $10\pi + \pi/2$ to find the optimum point, then approach it from the same direction as we did in the scan (to avoid hysteresis errors). This centres the beam on the ion within $0.5\,\mu\text{m}$.

To estimate the beam-pointing fluctuations from vibrations, and the drift in point-ing from mechanical relaxation, we set the polarisation of the $R_{\parallel}$ beam to give a large differential light-shift, and look at fluctuations in this light-shift. We align the beam carefully to find the peak light-shift, then misalign the beam horizontally to reduce the light-shift by a factor of 2. This greatly increases the sensitivity to beam pointing changes. As we know the beam-waist, we can translate the fluctuations in the light-shift to fluctuations in the beam position. Any beam pointing noise slow compared to the length of the Raman pulse ($\sim10\,\mu\text{s}$) will be apparent as excess shot-noise or drift on the light-shift. The results of this experiment are plotted in Fig. 6.10. Three things are obvious; there is excess shot-noise, there is a periodic fluctuation with the air-conditioning (8 minute period), and there is a linear trend of $0.4\,\mu\text{m/h}$. If we subtract off the linear trend and histogram the data the distribution is well described by a Gaussian with $\sigma = 570\,\text{nm}$. We expect the shot-noise to give $\sigma_{\text{stat}} = 200\,\text{nm}$. Subtracting these in quadrature we find an excess noise standard deviation of $\sigma_{\text{excess}} = 520\,\text{nm}$. The level of drift and noise we measure would give

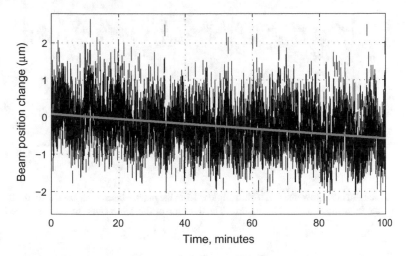

Fig. 6.10 Raman beam pointing noise. The beam waist is displaced by $\sim w_0$ horizontally to maximise sensitivity to position noise. The linear trend is $0.4\,\mu\text{m/h}$. The black bars are 1σ error bars. The excess noise (above shot-noise) is 520 nm

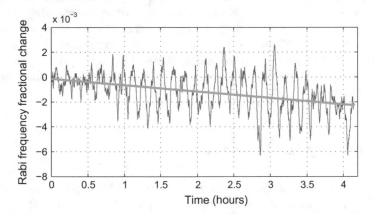

Fig. 6.11 R_V-$R_{\parallel}$ Raman Rabi frequency drift. At time zero the Raman beam pointing was optimised to give maximum Rabi frequency. The periodic fluctuations of $\sim 2 \times 10^{-3}$ are synchronous with the (8 min) air-conditioning cycle. The linear drift is $-5.1(2) \times 10^{-4}/\text{h}$. Each data point takes 1.2 s to acquire, and consists of $N = 200$ shots of the experiment. The data is filtered by a Savitzky-Golay filter with window length $n_r = n_l = 40$

average Rabi frequency errors of $<0.1\%$ if the beam was initially perfectly centred on the ion. A measurement of the pointing noise in the vertical direction gave similar results. As the other Raman beams are constructed out of optics mounted in the same way and have similar beam-path dimensions as the $R_{\parallel}$ beam we have characterized here, we expect similar results.

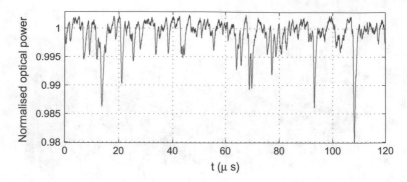

Fig. 6.12 Raman laser power noise. This sample is from the master laser, measured after the optical fibre with the intensity stabilisation running. The duration of the 'drop-outs' is $\sim 0.5\,\mu$s. The RMS noise is 0.2%

To characterize the long-term Rabi frequency drift we measure the Rabi frequency drift for the two beam-paths we use for the two-qubit gate of Chap. 8, R_V and $R_\parallel$. After aligning the Raman beams on the ion using the piezos as described, we apply a carrier pulse of area $10\pi + \pi/2$ and measure the drift in the resulting population over 4 h (Fig. 6.11). The linear drift in Rabi frequency is -5.1×10^{-4}/h. We also see fluctuations of 0.2% over the 8 minute air-condition cycle. Over the 4 h the measured Rabi frequency is always within 0.6% of the initial measurement. This Rabi frequency variation would cause an error of less than 10^{-4} on a single-qubit or two-qubit gate.

6.4.2   Intensity and Phase Noise

We start with the obvious test of measuring the Raman beam power noise after the optical fibre, with the intensity stabilisation running. This measurement (Fig. 6.12) shows that there is a significant level of 'fast' noise. However as our typical Raman pulse duration is several μs, these 'drop-outs' are averaged over, and hence do not cause significant errors – we model this carefully in Chap. 7.

To measure the potentially more troublesome slower noise we use an FFT analyser.[3] The noise spectra for both the master and slave laser are shown in Fig. 6.13. The peaking at ~ 1 kHz can be removed by increasing the Raman doubling cavity lock gains, at the cost of making the phase noise spectrum worse.

To estimate the phase noise we measure the close-in spectrum of the beat-note between the two Raman beams. We beat the two Raman beams onto a photo-diode[4] just before the beams enter the vacuum system. We then amplify the 3.2 GHz beat-

---

[3]SRS785: DC to 100 kHz FFT analyser.
[4]Newfocus 1437: 25 GHz bandwidth photo-diode.

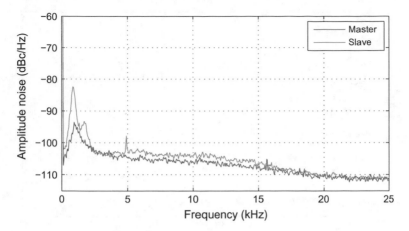

Fig. 6.13 Raman laser single-sided amplitude noise spectrum, measured with the intensity stabilisation running. The noise floor from the photo-diode and FFT analyser are far below the measured noise

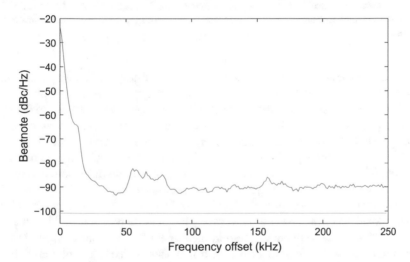

Fig. 6.14 Raman laser single-sided phase noise spectrum, measured between the master and slave laser. The *solid green line* is the signal chain noise floor. Further than ∼50 kHz from the carrier the noise appears white. This continues to the ∼1.5 MHz cavity bandwidth pole, after which the noise rolls off

note and either measure it directly with a microwave spectrum analyser or mix it down to ∼50 kHz and measure it with an FFT analyser. The measured spectrum contains both amplitude and phase noise, but from our AM noise measurements (Fig. 6.13) we are confident that these spectra are dominated by phase noise.

A wide-band measurement of the phase noise between the master and slave laser (Fig. 6.14) shows some close-in noise (<25 kHz) and a definite noise-floor. Addi-

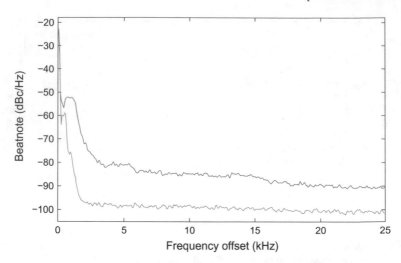

Fig. 6.15 Single-sided phase noise spectrum of the master laser versus the slave laser (*blue*) and the master laser versus itself (*green*). The measurement noise floor is approximately -100 dBc/Hz. The master/master spectrum has some noise close to the carrier (<1 kHz) but quickly reaches the noise floor

tional measurements show that, as expected, this floor drops off above the doubling cavity pole frequency of 1.5 MHz. A further measurement of the close-in phase noise both between the master and the slave laser and between two beams from the master laser (Fig. 6.15) shows several interesting features. The first is peaking in the noise at ~1 kHz. This peaking greatly increases as the doubling cavity lock gains are increased, causing significant single-qubit errors. The spectra here are measured after roughly optimising the lock parameters for good single-qubit gate fidelity.

The second interesting feature of these phase noise spectra is the large difference between the 'master-master' spectrum and the 'master-slave' spectrum – the 'master-slave' spectrum shows a significantly larger noise pedestal, and larger close-in noise. This is due to the large interferometer path for the 'master-slave' system, which includes the two doubling cavities. If this phase noise is a limitation in future experiments we could reduce it by phase-locking the beat-note between the master and slave laser before the AOM network, correcting the phase by feeding back to the master-slave injection locking AOM.

Finally we measure the very close-in phase noise between the Raman beams and clock qubit with a Ramsey experiment (Fig. 6.16). The fitted coherence time is $\tau = 3(1)$ s, compared to $\tau = 6(1)$ s measured between the microwave source and the clock qubit (Fig. 6.8). This suggests that the Raman beam phase noise is comparable with the qubit phase noise. This phase noise could be removed by the same phase feedback look as previously described, assuming that the short fibres after the AOM network do not add significant noise.

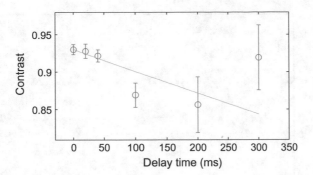

Fig. 6.16 Coherence between the co-propagating Raman beams and the clock qubit. The exponential fit gives a coherence time of $\tau = 3(1)$ s

6.5 Motion

6.5.1 Cooling

The equilibrium axial centre-of-mass mode temperature at the 'high fluorescence' Doppler-cooling parameters used for readout is $\bar{n} = 15 - 20$. Reducing the cooling beam intensity to roughly a saturation intensity and red-detuning to half-fluorescence greatly reduces the equilibrium temperature. Figure 6.17 shows the mode temperature as a function of this 'low Doppler' cooling time. We find a cooling rate of $\tau = 0.46(5)$ ms and an equilibrium temperature of $\bar{n} = 6.4(2)$. For a two-level system the minimum mode temperature possible ($\delta = -\Gamma/2, I \rightarrow 0$) with our cooling beam geometry ($45°$ projection onto $\hat{z}$) is $\bar{n} = 4.6$. As our system is nowhere near a two-level system we are getting respectably close to this limit.

After low-Doppler cooling we switch to continuous sideband cooling. This involves driving the red sideband of the 'stretch' qubit while simultaneously repumping with the 397σ beam. Figure 6.18 show the mode temperature versus continuous sideband cooling duration. After the continuous cooling has reduced the mode temperature to below $\bar{n} = 0.5$ we move to pulsed sideband cooling. We use an intermediate stage of continuous cooling as the achieved cooling rate is significantly higher than with pulsed sideband cooling. For a fixed cooling time this allows us to reclaim more population from high lying motional states, improving the final temperature. For one or two ions we reliably cool all axial modes to $\bar{n} \sim 0.01$.

6.5.2 Motional Frequency Drifts

We measure the axial frequency by scanning the frequency of a weak 'tickle' drive added onto one of the trap end-caps and monitoring the fluorescence for a single

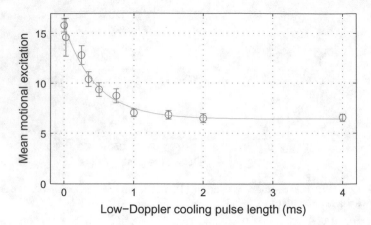

Fig. 6.17 Single ion axial motional temperature versus 'Low-Doppler cooling' duration. The ion starts at $\bar{n} = 15$ – the equilibrium temperature at the 'high fluorescence' parameters. The cooling rate at the 'low-Doppler' parameters is $\tau = 0.46(5)$ ms, with an equilibrium temperature of $\bar{n} = 6.4(2)$

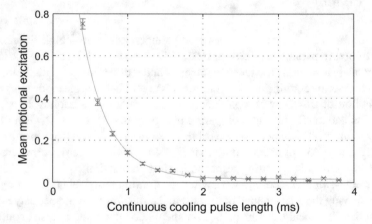

Fig. 6.18 Mode temperature as a function of continuous sideband cooling pulse length. The continuous cooling pulse is applied after 1 ms of 'low Doppler' cooling. The cooling rate is $\tau = 360(20)$ μs and the equilibrium temperature $\bar{n} = 0.017(2)$. The 397σ repumping rate is $\tau = 5.3(2)$ μs and the Raman sideband Rabi π-time $t_\pi = 23$ μs

ion. For weak Doppler cooling (typically the Doppler cooling beam is red detuned give 5% of peak fluorescence) we see increased fluorescence when the tickle drive is resonant with the motion. With the tickle amplitude optimised (to give a 20% increase in fluorescence) we measure a Lorentzian line with $\approx$80 Hz full width at half maximum.

We monitor the drift of the axial frequency by repeatedly scanning over the resonance. Figure 6.19 shows a typical result. We see modulation of the axial trap frequency from the air conditioning on top of a slow drift. However we also infrequently

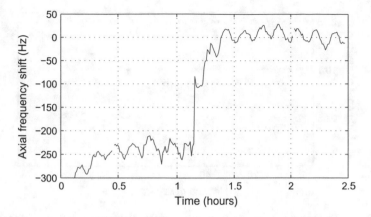

Fig. 6.19 Axial frequency drift. The periodic fluctuations have a peak-peak amplitude of 30 Hz and are synchronous with the air conditioning cycle. This data set captures an infrequent jump in the axial frequency. Each frequency measurement takes 35 s, during which we scan the tickle frequency over 1 kHz in 10 Hz steps. The missing data points are due to poor fits

see much larger jumps. The cause of these are unknown, but we suspect it is due to mechanical shifts in the helical resonator leading to changes in the RF voltage, which modulate the axial trap frequency due to the axial micromotion; we have also seen such jumps in the power reflected from the helical resonator. The 250 Hz axial frequency change would correspond to a ∼10 kHz radial frequency shift which is broadly consistent with the level of drift that we have observed.

6.5.3 Motional Heating

To measure the ambient heating rate we cool the ion to near the ground state, then turn off all laser beams to let the ion heat undisturbed for a period of time. Finally we measure the temperature of the ion with a red and blue sideband pulse. Repeating this for several different delays allows us to map temperature versus heating time (Fig. 6.20). A linear fit gives the heating rate of the $f_z = 2.047$ MHz axial mode of a single ^{43}Ca$^+$ ion as $\dot{\bar{n}} = 1.1(1)$s^{-1}. This corresponds to an electric-field noise power spectral density of $S_E = 1.6 \times 10^{-14}$V^{2}m^{-2}Hz^{-1}. This is comparable to or better than other traps of similar size [BKRB14].

6.5.4 Motional Coherence

We measure the coherence time of the motion by performing a Ramsey experiment between the motional states $|n = 0\rangle$ and $|n = 1\rangle$. Any noise in the trap voltages that

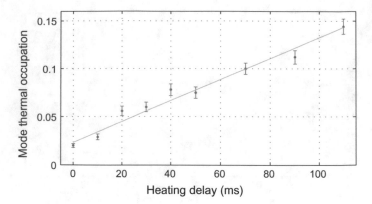

Fig. 6.20 Ambient heating of the axial motion for a single ^{43}Ca$^+$ ion with $f_z = 2.047$ MHz. The linear fit gives a heating rate of $\dot{\bar{n}} = 1.1(1)\text{s}^{-1}$

Fig. 6.21 Motional Ramsey experiment. The fringe contrast is a measure of $|\rho_{01}|$, the coherence between motional states $|n = 0\rangle$ and $|n = 1\rangle$. The *green line* is a fit to an exponential decay giving a coherence time $\tau = 190(30)$ ms

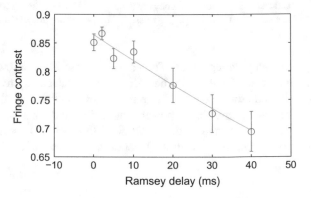

is adiabatic compared to the motional frequency will not cause motional transitions (changes in the motional populations), but will scramble the phase of the coherences between states.

To perform this motional Ramsey experiment we prepare the state $|\downarrow, n = 0\rangle + |\downarrow, n = 1\rangle$ using a carrier $\pi/2$ followed by a red sideband π-pulse. After a period of free evolution, during which we are sensitive to motional dephasing, but insensitive to (spin) qubit dephasing, we map back to the spin qubit with another red sideband π-pulse and a carrier $\pi/2$. Scanning the phase of the final $\pi/2$ allows us to measure the phase and amplitude of ρ_{01} – the coherence between motional states $|0\rangle$ and $|1\rangle$. We define the motional coherence time τ as the decay time of the coherence between $|n = 0\rangle$ and $|n = 1\rangle$ (see Sect. 4.4.6).

An experimental dataset is shown in Fig. 6.21. The initial contrast is below 1 due to an incorrectly set red sideband pulse area. We find a coherence time of $\tau = 190(30)$ ms. For delay times larger than ~50 ms we see significantly non-statistical data due to slow motional frequency drifts.

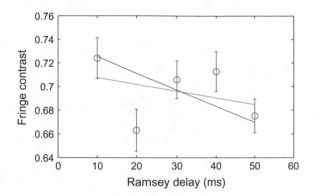

Fig. 6.22 Motional spin-echo experiment. The motional state $|0\rangle + |1\rangle$ is allowed to evolve for $t/2$, the states are then flipped, and the state evolved for another $t/2$. This removes any phases caused by motional frequency errors constant over the length of the sequence. The *green line* is a fit to an exponential decay, giving $\tau = 1(2)$ s. The red line is a fit with a fixed coherence time of $\tau = 0.5$ s

To try to separate the drifts in motional frequency from true dephasing we perform a motional spin echo experiment (Fig. 6.22). This is similar to the motional Ramsey experiment, but with a red sideband π-pulse, carrier π-pulse, and red sideband π-pulse in the middle of the free evolution. Thus we measure the difference in accumulated motional phase in the first arm of length $t/2$ to the second arm of $t/2$. Any offsets that are constant over the course of the sequence, such as slow motional frequency drifts, are removed. We measure a coherence time $\tau \sim 1$ s. Following [TMK+00] we find that the heating alone causes a dephasing at a rate $\tau = (2\dot{\bar{n}})^{-1} \sim 0.5$ s. The red line on Fig. 6.22 is a fit with the coherence time fixed at this value – the coherence time we measure with this spin-echo sequence is consistent with the limit from heating.

References

[BMB+98]   Berkeland, D.J., J.D. Miller, J.C. Bergquist, W.M. Itano, and D.J. Wineland. 1998. Minimization of ion micromotion in a Paul trap. *Journal of Applied Physics* 83 (10).

[BKRB14]   Brownnutt, M., M. Kumph, P. Rabl, and R. Blatt. 2014. Ion-trap measurements of electric-field noise near surfaces. arXiv:1409.6572.

[Hom06]   Home, J.P. 2006. *Entanglement of Two Trapped-Ion Spin Qubits*. Ph.D thesis, University of Oxford.

[McD03]   McDonnell, M.J. 2003. *Two-Photon Readout Methods for an Ion Trap Quantum Information Processor*. Ph.D thesis, University of Oxford.

[MSW+08]   Myerson, A., D.J. Szwer, S.C. Webster, D.T.C. Allcock, M.J. Curtis, G. Imreh, J.A. Sherman, D.N. Stacey, A.M. Steane, and D.M. Lucas. 2008. High-fidelity readout of trapped-ion qubits. *Physical Review Letters* 100 (20): 200502.

[Szw09]   Szwer, D.J. 2009. *High Fidelity Readout and Protection of a 43Ca+ Trapped Ion Qubit*. Ph.D thesis, University of Oxford.

[TMK+00]   Turchette, Q.A., C.J. Myatt, B.E. King, C.A. Sackett, D. Kielpinski, W.M. Itano, C. Monroe, and D.J. Wineland. 2000. Decoherence and decay of motional quantum states of a trapped atom coupled to engineered reservoirs. *Physical Review A* 62 (5): 1–22.

Chapter 7
Single-Qubit Gates

In this chapter we discuss the implementation and benchmarking of laser driven single-qubit gates. We discuss the sources of error, and determine what experimentally limits our gate fidelity.

7.1 Randomized Benchmarking of a Single Qubit

In this chapter we perform single-qubit randomized benchmarking using a minor modification of the Knill algorithm [KLR+08]. We summarise the algorithm here.

We generate a randomized sequence of l 'computational gates' from a uniform sampling of $\{\pm x, \pm y\}\,\pi/2$ pulses. Before each random $\pi/2$ pulses we insert π pulses about an axis randomly selected from the set $\{\pm x, \pm y, \pm z, \pm I\}$. The final part of the sequence consists of another random π pulse, and one of the two (randomly chosen) $\pi/2$ pulses that brings the system back into the σ_z basis. Finally, we apply another random π pulse. The first l pairs of π and $\pi/2$ pulses comprise the sequence under test – the final 3 pulses are a randomized measurement, the error of which is absorbed into the state preparation and measurement constant error. To benchmark our gate implementation we generate N_G random sequences for each length l. For each different sequence we repeat the experiment N_e times. By comparing the experiment measurements to the result we calculate that each sequence should give, we calculate the probability of error. Assuming sufficient randomisation, the probability of error, p_l, for a sequence of length l is described by

$$p_l = \frac{1}{2}\left[1 - (1 - d_{if})(1 - d)^l\right] \approx \frac{1}{2}(d_{if} + ld) \tag{7.1}$$

© Springer International Publishing AG 2017
C.J. Ballance, *High-Fidelity Quantum Logic in Ca+*,
Springer Theses, DOI 10.1007/978-3-319-68216-7_7

where $d_{if} = 2\epsilon_{SPAM}$ is twice the constant state-preparation and measurement (SPAM) error, d is the depolarisation rate per gate, and the approximation holds for small d, d_{if}. The error (infidelity) per gate is given by $\epsilon = 1 - F = d/2$.

In our experiment we implement only $\pi/2$ pulses. We form the π pulses from two identical $\pi/2$ pulses in series. We implement identity pulses by doing nothing for the length of a pulse. We implement σ_z pulses by performing an identity pulse and shifting the phase of all subsequent pulses in software. On average, we apply pulses of area $\pi/2$ for each π-pulse, as 50% of the π pulses are about the $\{\pm I, \pm Z\}$ axes. Thus each 'randomized computational gate', consisting of a $\pi, \pi/2$ pair has an average applied pulse area of π.

For the work in this chapter we use $N_G = 32$ random sequences for each sequence length l. We repeat each sequence $N_e = 300$ times. The sequences are pre-calculated, and all experiments with the same l use the same set of (pseudo-) random sequences.

7.2 Experimental Setup

We implement the randomised benchmarking on the low-field clock qubit ($S_{1/2}^{4,+0} \leftrightarrow S_{1/2}^{3,+0}$). We drive the qubit transition by a pair of Raman beams co-propagating along the magnetic field direction. For this geometry the Lamb-Dicke parameter is $\eta \approx 10^{-6}$. This means that the Raman interaction is insensitive to the ion's motion. The two beams are linearly cross-polarised, such that the polarisation vectors in the spherical basis are $\mathbf{e_r} = \{1, 0, 1\}/\sqrt{2}$ and $\mathbf{e_b} = \{1, 0, -1\}/\sqrt{2}$. This maximises the Rabi frequency (see Table 3.2).

For simplicity we state-prepare by using a π optical pumping beam on the 397 transition (rather than using a series of microwave pulses starting from the $S_{1/2}^{4,+4}$ state). Due to polarisation impurities and off-resonant pathways that limit the optical pumping quality our total state-preparation and measurement (SPAM) error is $\approx 3\%$. This is not a limitation for these experiments, as we can always use long enough sequences of gates that the gate error dominates over the SPAM error.

We use the magnetic field servo to fix the frequency of the stretch qubit to better than 1 kHz. The linear sensitivity of the clock qubit at our operating field of 1.93 G is $\frac{df}{dB} \approx 4.7\,\mathrm{Hz\,mG^{-1}}$, thus our servo fixes the clock qubit frequency to less than 1 Hz. We find the clock qubit frequency using a Ramsey experiment to a statistical uncertainty of $< 1\mathrm{Hz}$.

The benchmarking pulse sequence is generated by an FPGA with a timing resolution of 10 ns. The 'dead' time between pulses is typically 0.5 μs. This time is used to select the next pulse phase. The acoustic delay in the AOMs used to switch the two Raman beams is ~1 μs. We adjust the timing of the switch signals so as to align the two optical pulse edges to better than 100 ns – this minimises excess photon scattering error. (We note that for the shortest $\pi/2$ pulses we use, 900 ns, the DDS sources have already selected the next pulse phase by the time the previous optical pulse arrives at the ion!) Off-resonant excitation is not an issue in these experiments (see Sect. 7.3.2 below), so we do not attempt to control the pulse shape. We observe a typical 10–90% rise time of 100 ns on both beams.

We calibrate the $\pi/2$ pulse length by performing (typically) 21 identical pulses with the same timing as for a benchmarking sequence. We scan the pulse length over ~10% and fit the resulting fringes. This determines the correct pulse length with an uncertainty of ~10^{-3}.

7.3 Error Sources

In this section we discuss the sources of error that we might expect to limit our gate fidelity.

7.3.1 Photon Scattering

Ignoring any positive fidelity contribution from the state after scattering, and ignoring scattering into the D states (i.e. following [OIB+07]), the error from scattering for a π-pulse is

$$\epsilon = \frac{2\pi}{3} \frac{\gamma \omega_f}{\Delta(\Delta - \omega_f)} \tag{7.2}$$

For the two detunings used in this chapter, $\Delta = \{-1.01, -1.91\}$ THz, this scattering error is $\epsilon = \{3.8, 1.8\} \times 10^{-5}$. Using the full 5 mW in each beam gives $t_\pi = \{1.6, 3.5\}\,\mu$s. As each randomised gate, consisting of a random-axis π-pulse and a random-axis $\pi/2$-pulse, has an average Raman pulse area of π (as the identity and Z pulses are implemented in software) the scattering error per randomised gate is the same as the scattering error per Raman π-pulse.

7.3.2 Off-Resonant Excitation

The low-field clock qubit transition has strongly suppressed off-resonant transitions assuming the Raman beam polarisations are perfect (Table 3.2). For a carrier $t_{\pi/2} = 0.5\,\mu$s and $\Delta = -1$ THz, these $\delta m_f = 2$ transitions give an off-resonant excitation error of ~10^{-8}. A larger error comes from the possible polarisation impurity in the Raman beams: a π component of 1% amplitude gives an error ~10^{-5} for $t_{\pi/2} = 0.5\,\mu$s. Such a polarisation impurity could come from a $0.6°$ misalignment of the beam with the magnetic field – this is roughly the level to which we align it. As this off-resonant error decreases quadratically with the gate time we expect it to be negligible for these experiments (with $1\,\mu s \lesssim t_{\pi/2} \lesssim 100\,\mu s$).

Fig. 7.1 Benchmarked
error-per-gate versus
detuning. We detune the
local oscillator from the
qubit frequency and measure
the error-per-gate with
$t_\pi/2 = 1.78\,\mu s$ and
$t_{dead} = 2.35\,\mu s$ (blue *points*).
The *solid line* is a model with
no free parameters, except
for a vertical offset added to
account for other sources of
error (color figure online)

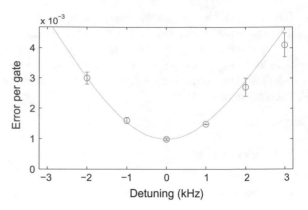

7.3.3 *Detuning*

Systematic detuning errors can cause significant errors. However for our relatively
high Rabi frequencies, small dead times, and frequency-stable qubit these errors
should be negligible. The slowest gates we implement have $t_\pi = 100\,\mu s$, for which
a 10 Hz detuning gives a 10^{-5} error-per-gate. We expect our qubit frequency to be
controlled better than 10 Hz.

To confirm our model for detuning errors we measure the error-per-gate while
varying the detuning from the qubit (Fig. 7.1). The data are in good agreement with
the model.

An additional source of error is the unavoidable light-shift from the Raman beams.
This light-shift is proportional to the Raman Rabi frequency, hence the Raman detun-
ing alone sets the error magnitude. For a detuning of $\Delta = -1\,\text{THz}\ \Delta_{LS}/\Omega = 0.004$,
leading to an error-per-gate of $\sim 10^{-5}$. This error can be mostly removed by tuning
into resonance with the light-shifted qubit, but we do not do this here as the error is
negligible compared to other errors.

7.3.4 *Systematic Rabi Frequency Error*

A systematic pulse area fractional error of 0.25% gives an error-per-gate of 1×10^{-5}.
The measured pointing stability of the Raman beams gives a fractional Rabi frequency
drift of $\sim 0.5 \times 10^{-3}$/h (Sect. 6.4.1), thus long-term drift should not be an issue, as
randomized benchmarking runs take $\sim 10\,\text{min}$ to perform.

To confirm our model, we measure the error-per-gate while varying the pulse
length (Fig. 7.2). The data are in good agreement with the model.

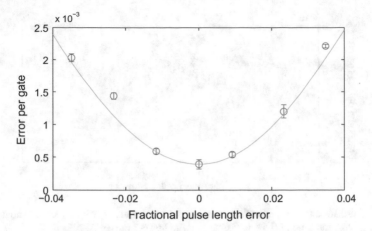

Fig. 7.2 Benchmarked error-per-gate versus pulse length error. We vary the $\pi/2$ pulse length about the optimum of $4.30\,\mu s$. The *solid line* is a model with no free parameters, except for a vertical offset added to account for other sources of error

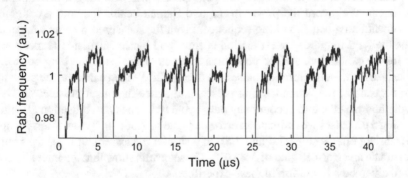

Fig. 7.3 Sample of Raman Rabi frequency noise calculated from the measured Raman beam powers during a sequence. The bandwidth of the photo-diodes used to measure the Raman beam powers was $\approx 20\,MHz$. Power transients during each pulse and $\sim 1\%$ 'dropouts' are visible

7.3.5 Rabi Frequency Noise

Looking at a sample of intensity noise (Fig. 7.3) it is clear that this may be a significant source of error in this experiment. As well as noise in the output power from the doubling cavities, there are power transients at the start of a sequence of pulses due to fast thermal effects in the AOM. To characterize the effect of the noise we could sample the power noise in each laser beam in a continuous fashion, then integrate from t_0 to $t_0 + \tau$ (where τ is our pulse length) for many randomly chosen t_0 to get an idea of the variance of our pulse area. This would likely give an overly optimistic estimate as it ignores duty-cycle effects.

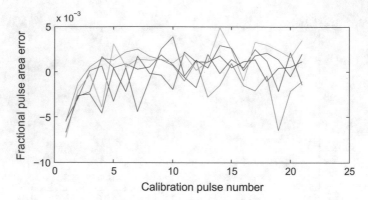

Fig. 7.4 Pulse areas for 5 repeats of a 21 pulse calibration sequence. We set the mean area of these pulses to $\pi/2$. A transient of 0.5% is visible over the first $\sim$3 pulses

A fairer method is to measure the pulse areas while running a benchmarking experiment. We record the power in the two Raman beams for both a calibration experiment and a benchmarking sequence. From the measured powers we can calculate the Rabi frequency as a function of time, and hence calculate the mean pulse area for the calibration, and the pulse area for each of the benchmarking pulses.

We take the mean calibration pulse area to be $\pi/2$. We then calculate the pulse area error for each of the benchmarking pulses, and hence the mean pulse error. Using this method we take into account any duty cycle or intensity stabilisation effects.

Using 5 μs pulses we estimate an error-per-gate of 7.5×10^{-6} from intensity noise and transient effects (Figs. 7.4 and 7.5). In a similar measurement for 0.5 μs pulses we estimate an error-per-gate of 5×10^{-5}. These results show that the intensity noise is unlikely to be a limitation in these experiments.

7.3.6   Phase Noise

There are two dephasing mechanisms; magnetic field noise modulating the qubit frequency, and noise in the Raman beam difference phase. As we measure the 'clock' qubit coherence time to be long compared to any sequence (Sect. 6.3.3) we expect laser noise to dominate.

Broadly speaking, we expect to see two kinds of laser phase noise: a white(ish) noise floor and a slow phase wander. Noise that is white compared to the pulse bandwidth causes a larger error for shorter pulse lengths, due to the pulse encompassing a larger noise bandwidth. For pure white noise the average π-pulse fidelity can be analytically calculated to be [CBWT12, Har13]

$$\epsilon = \frac{\pi\Omega}{6}\Gamma \tag{7.3}$$

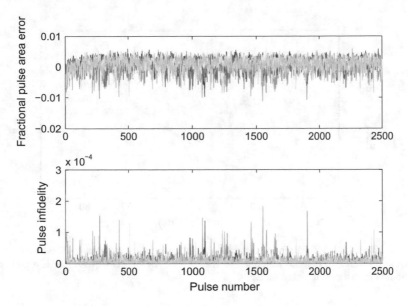

Fig. 7.5 Pulse area error and inferred pulse infidelity for 5 repeats of a sequence of 2500 pulses. The mean error-per-gate is 7.5×10^{-6}

where Γ is the single-side power spectral density in (fractional) power per Hz.

For a slow phase wander individual pulses can be considered perfect, but the relative phase of the pulses drifts. This error increases for longer pulse lengths. Consider a small periodic phase modulation of $3°$ at $20 \, \text{kHz}$ (perhaps caused by a $\sim$ nm acoustic vibration of a mirror). This is unobservable on a Ramsey experiment: it causes at most a 10^{-3} contrast error which does not increase with the Ramsey delay. This modulation would cause an error-per-gate of 10^{-3} – much larger than the other terms in our error budget.

Our broad-band measurement of the phase noise between the two beams used to drive the qubit Raman transitions (Master and Slave, Fig. 6.14) show that above $50 \, \text{kHz}$ there is a $-90 \, \text{dBc/Hz}$ white noise floor. Above the frequency-doubling cavity bandwidth of $\sim 1.5 \, \text{MHz}$ this drops off. Closer to the carrier ($< 5 \, \text{kHz}$) the noise increases substantially (Fig. 6.15) – as this is much slower than our pulse bandwidth this will give a phase wander error.

Applying (7.3) we find the error from the white noise to be $\epsilon = 8 \times 10^{-4}$ for $t_\pi = 2 \, \mu\text{s}$, assuming the white noise floor extends far above $1/t_\pi$. As the white noise drops off at frequencies above the doubling cavity bandwidth this is a slight overestimate. This white noise will be a substantial source of error for fast gates.

We estimate the error from the slow phase wander from a numerical simulation. We create random phase trajectories that have a noise spectrum identical to our measured low-frequency phase noise (Fig. 6.15). We then simulate a benchmarking experiment where each pulse is perfect, but the phase between each pulse varies from the ideal. We find the error for a given pulse length by randomising over many phase

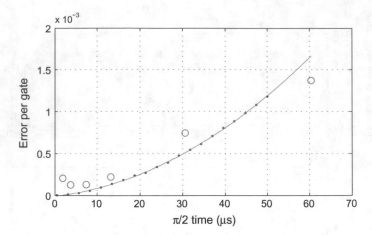

Fig. 7.6 Phase noise simulation (*blue points*), based on the measured Master/Slave phase noise spectrum (Fig. 6.15), alongside experimental results (*red circles*). The *green line* is a 2nd order polynomial fit to the numerical simulation

trajectories. The numerical results are the blue points in Fig. 7.6. After measuring the phase noise we experimentally measured the error-per-gate for different pulse lengths – these results are the red points in Fig. 7.6. There is a good agreement between the experimental results and our predicted error from phase noise: this is thus the dominant error as we increase the gate length.

We minimise this phase noise error by optimising the Raman laser doubling cavity lock parameters for both lasers. Using a 60 μs gate time and a 50 gate sequence, we adjust the lock to minimise the sequence error. Initially the error-per-gate was $\sim 15 \times 10^{-4}$. Setting P (proportional lock path gain) and D (derivative gain) to zero, and increasing I (integral gain) to just below loop oscillation increases the error to $\sim 50 \times 10^{-4}$ per gate. The minimum error occurs when P and D are zero, and the I gain is set approximately half-way between the lock 'dropping out' and oscillating. This gives an error of $\sim 6 \times 10^{-4}$.

7.4   Results

Our error analysis suggests that the major limit in our gate fidelity will be phase noise, rather than photon scattering. We only use two values of Raman laser detuning (which sets the photon scattering error-per-gate), and vary the Raman beam power to vary the gate speed. Figure 7.7 is an example of the data we fit to get the error-per-gate for each pulse length. A summary of all experimental results is shown in Fig. 7.8.

The gate fidelity is limited by laser phase noise for both slower than optimum and faster than optimum gates. The lowest error-per-gate is $6.6(3) \times 10^{-5}$ for $\pi/2$-pulse length 7.45 μs. The error from photon scattering for this point is 1.8×10^{-5}.

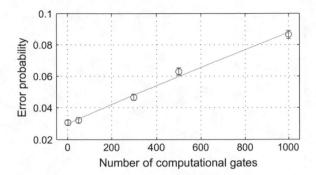

Fig. 7.7 A typical randomised benchmarking experiment dataset for the lowest gate error achieved. The $\pi/2$ pulse length is 7.45 µs. The fitted curve gives $\epsilon_{SPAM} = 3.0(1) \times 10^{-2}$ and $\epsilon_{PG} = 6.6(3) \times 10^{-5}$. The error bars on the data points are statistical

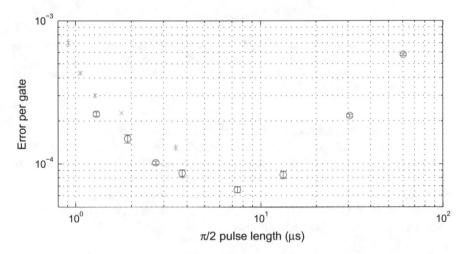

Fig. 7.8 Single-qubit benchmarking results. The *blue circles* are taken with $\Delta = -1.91$ THz, and the *red crosses* with $\Delta = -1.01$ THz. The difference in scattering rate between the two detunings gives a difference in EPG of 2×10^{-5} – much smaller than the observed difference. We believe that this is due to the laser phase noise spectrum changing after adjusting the Raman detuning (and hence realigning the doubling cavities)

Extrapolating the error from the fast phase noise we predict $\sim 4 \times 10^{-5}$ error for this point. We expect the systematic errors in pulse area and detuning to give errors below 10^{-5}.

For the shortest gates at $\Delta = -1.01$ THz detuning the error appears to increase faster than the $1/t_\pi$ we expect from white phase noise. This is possibly due to finite timing resolution in setting the pulse length. For the shortest gate (pulse length 0.91 µs) we expect $> 10^{-4}$ error-per-gate from this systematic.

As a comparison, we also benchmark our single-qubit microwave gates. For microwave gates the sources of error change considerably. The phase noise of the

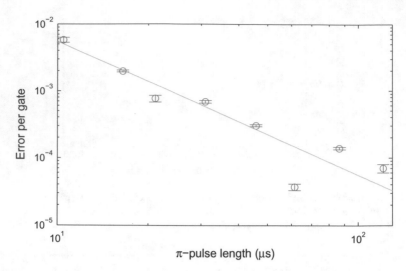

Fig. 7.9 Benchmarking of microwave driven single-qubit gates. The dominant error is off-resonant excitation. The *solid line* is a quadratic fit to guide the eye, and agrees roughly with the error predicted from the measured microwave polarisation. The scatter in the data is due to the unresolved $\sin^2$ structure of the off-resonant excitation

microwave source is negligible (giving $\lesssim 10^{-8}$ error-per-gate [Har13]), and there is no photon scattering. However there is a large off-resonant excitation error due to the uncontrolled microwave polarisation driving the neighbouring transitions (only ≈ 700 kHz off-resonance). The experimental results (Fig. 7.9) show how this off-resonant excitation dominates all other errors. For the slowest gate ($t_\pi = 121\,\mu$s) the error-per-gate is $7(1) \times 10^{-5}$. This again confirms that the large errors for comparable length laser gates is due to laser noise, not qubit dephasing.

References

[KLR+08]   Knill, E., D. Leibfried, R. Reichle, J.W. Britton, R.B. Blakestad, J.D. Jost, C.E. Langer, R. Ozeri, S. Seidelin, and D.J. Wineland. 2008. Randomized benchmarking of quantum gates. *Physical Review A* 77 (1): 1–7.

[OIB+07]   Ozeri, R., W.M. Itano, R.B. Blakestad, J.W. Britton, J. Chiaverini, J.D. Jost, C.E. Langer, D. Leibfried, R. Reichle, S. Seidelin, J.H. Wesenberg, and D.J. Wineland. 2007. Errors in trapped-ion quantum gates due to spontaneous photon scattering. *Physical Review A* 75 (4): 1–14.

[CBWT12] Chen, Z., J.G. Bohnet, J.M. Weiner, and J.K. Thompson. 2012. General formalism for evaluating the impact of phase noise on Bloch vector rotations. *Physical Review A* 86 (3): 032313.

[Har13]   Harty, T.P. 2013. *High-Fidelity Microwave-Driven Quantum Logic in Intermediate-Field 43Ca+*. Ph.D thesis.

Chapter 8
Experimental Implementations of Two-Qubit Gates

In this chapter we present the results of our two-qubit gate experiments. We start by discussing how we define gate fidelity and the experimental signals we measure. We present preliminary experiments of a light-shift gate on a pair of ^{40}Ca$^+$ qubits. We then present in detail the experimental setup and results of a light-shift two-qubit gate on two ^{43}Ca$^+$ qubits, including an error budget that gives good agreement with experiment over two orders of magnitude in gate speed. Finally, we present a modification of the light-shift gate applied to entangle two different atomic species, ^{40}Ca$^+$ and ^{43}Ca$^+$, and describe how the same method could be used to entangle Ca$^+$ and Sr$^+$.

8.1 Measuring Gate Fidelity

How can we measure the fidelity of an experimental implementation of a two-qubit gate? Rather than performing the 'honest' experimental test of measuring the output state fidelity for a complete basis of input states (process tomography) it is common to 'cheat'. The problem with process tomography is that it is difficult to make the required input states and analyse the generated output states with a high enough level of accuracy that the two-qubit gate error is dominant. In almost all experiments to date the cheat involves applying the gate to a single separable input state to generate a maximally entangled state, then analysing the fidelity of the entangled state. The assumption made here is that the gate errors are well enough understood that we can be confident that the error we measure for the one input state we use is typical of all input states. As we discussed in Sect. 4.4 this is generally true. The main disadvantage of this method (and even more so for process tomography) is that to get a small statistical uncertainty on the gate error a large amount of data needs to be taken.

The remaining methods involve applying multiple gates to amplify the error. The simpler method involves applying the same gate operation multiple times to one initial state. This certainly amplifies the gate error, but coherent gate errors (for

© Springer International Publishing AG 2017
C.J. Ballance, *High-Fidelity Quantum Logic in Ca$^+$*,
Springer Theses, DOI 10.1007/978-3-319-68216-7_8

example, an error in the gate geometric phase) grow quadratically in the number of gates, whereas if the gates are about random axes (as in a computational context) these errors are linear in the number of gates. The solution is to measure the error of a combination of many single-qubit gates and two-qubit gates using randomised benchmarking [GMT+12], and then to measure the additional error from interleaving further two-qubit gates. This measures the gate fidelity averaged over all initial states in a very thorough way. The difficulty with this technique is achieving small enough (addressed) single-qubit gate errors that they are comparable to or smaller than the two-qubit gate error.

In the experiments described in this chapter we use the two-qubit gate to generate the Bell state $|00\rangle + |11\rangle$, and measure the fidelity of this Bell state. If we have a $\sigma_x\sigma_x$ gate we can produce the Bell state directly from our initial state $|00\rangle$. For a $\sigma_z\sigma_z$ gate we need additional single-qubit gates to produce the Bell state. In both of these cases we need to know the state-preparation and measurement (SPAM) errors in order to separate the gate error from the less interesting errors. For the $\sigma_z\sigma_z$ gate we also need to know the error contribution from the single-qubit operations.

The problem of measuring our two-qubit gate error is now translated to measuring the fidelity with which we produce the Bell state $|\Phi\rangle := \frac{1}{\sqrt{2}}(|00\rangle + e^{i\phi_0}|11\rangle)$. We perform a reduced tomography procedure that allows us to measure efficiently the fidelity of our state with respect to the Bell state. This is a commonly used technique [BIWH96, SKK+00] which we describe here for completeness. This method scales easily to an N-qubit GHZ state [Mon11]. Calculating the fidelity of an arbitrary mixed state ρ to the Bell state gives

$$\mathcal{F} = \langle\Phi|\,\rho\,|\Phi\rangle = \frac{1}{2}(\rho_{00,00} + \rho_{11,11}) + \frac{1}{2}(e^{i\phi_0}\rho_{11,00} + \text{c.c.}) \qquad (8.1)$$

Measuring $\rho_{00,00} + \rho_{11,11} = P_{11} + P_{00}$ is simple – we just projectively measure the state. To measure the 'two-ion coherence' $\rho_{11,00}$ we can apply a single-qubit rotation $R(\pi/2, \phi)$ to both ions (the 'analysis' pulse) and measure the parity $\mathcal{P} := \langle\sigma_z \otimes \sigma_z\rangle$:

$$\mathcal{P} = P_{11} + P_{00} - P_{10} - P_{01} =$$
$$(1 - \rho_{11,11} - \rho_{00,00} - \rho_{10,10} - \rho_{01,01} + \rho_{10,01} + \rho_{01,10}) + (e^{-2i\phi}\rho_{11,00} + \text{c.c.}) \qquad (8.2)$$

By measuring this signal as a function of ϕ we can extract the phase and amplitude of the two-ion coherence, and from this calculate the fidelity.

8.2 ^{40}Ca$^+$-^{40}Ca$^+$ Light-Shift Gate

In our preliminary experiments we implemented the light-shift gate on a pair of ^{40}Ca$^+$ ground level qubits. The intent of this was to compare the fidelity reached in the new generation of the experiment to the previous generation [HML+06], which

Fig. 8.1 Parity signal versus analysis pulse phase ϕ for the state produced using a 57 μs ^{40}Ca$^+$-^{40}Ca$^+$ light-shift gate with Raman detuning $\Delta = -450$ GHz. The readout errors are *not* normalised out. After normalising out the readout errors this corresponds to a $\mathcal{F} = 1.005(7)$ Bell state

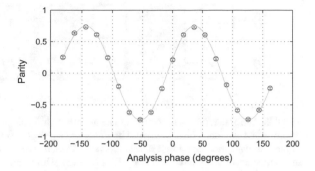

reached $\mathcal{F} = 83(2)\%$ (after normalising out readout errors). Using an EIT state-selective shelving process followed by manifold-selective fluorescence [MSW+04] we obtain typical readout errors of 5 and 90% for the two spin-states of each qubit. A representative plot of the parity signal produced by analysing the Bell state is given in Fig. 8.1. Without normalising out the readout errors the state fidelity is $\mathcal{F} \approx 80\%$. After normalising out the readout errors we find $\mathcal{F} = 1.005(7)$, with $P_{00} + P_{11} = 1.007(10)$ and $2|\rho_{11,00}| = 1.002(9)$. This is broadly consistent with the $\approx 5 \times 10^{-3}$ error we expect. Normalising out readout errors this large is fraught with danger; the EIT readout process depends sensitively on laser detuning and intensity, and unequal axial micro-motion or unequal illumination of the ions by the readout beams can cause underestimates of the readout errors and hence overestimates of the state fidelity.[1] This experiment shows that we have removed all the significant sources of errors that hindered the previous generation of experiment, but to measure the gate error with higher accuracy we clearly need much lower readout errors and a readout process that is more robust.

8.3 ^{43}Ca$^+$-^{43}Ca$^+$ Light-Shift Gate and Error Budget

In this section we present the results of a thorough experimental investigation into the sources of error in a light-shift gate on the stretch qubit ($S_{1/2}^{4,+4} \leftrightarrow S_{1/2}^{3,+3}$) of ^{43}Ca$^+$. We describe in detail how the experiment was implemented, and discuss the analysis of the various sources of error.

8.3.1 Experimental Setup

We perform a two-loop gate embedded in a spin-echo sequence, with one half of the gate in each arm of the spin-echo (see Fig. 8.2). The spin-echo pulses and the analysis

---

[1] My personal best is $\mathcal{F} = 103\%$.

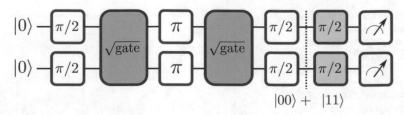

$$|00\rangle + |11\rangle$$

Fig. 8.2 The gate experiment sequence. The two light-shift gate pulses (*blue boxes*) are inserted in the arms of a spin-echo (*white boxes*). The π- and $\pi/2$-pulses are driven by microwaves. At the end of the spin-echo sequence a Bell state is produced (*dotted vertical line*). We either directly measure this Bell state, or apply a further 'analysis' $\pi/2$-pulse with a variable phase (*green boxes*)

pulse are driven by microwaves to decouple the quality of the spin-echo from the Raman beam power and detuning (which we may want to vary independently). We use a spin-echo sequence rather than the simpler option of a pair of $\pi/2$-pulses to reduce the sensitivity to qubit frequency errors. The stretch qubit is first-order magnetically sensitive, and magnetic field noise causes non-negligible spin dephasing over the typical gate durations we use. As the light-shift gate Hamiltonian commutes with σ_z this qubit frequency noise does not affect the gate operation. We split the gate into two halves for two reasons; firstly any small 'single beam' differential light-shifts cancel out, and secondly this keeps the total spin-echo length shorter, reducing spin dephasing errors. A typical gate experiment consists of ground-state cooling both axial motional modes (to $\bar{n} < 0.05$), state preparation by optical pumping, the spin-echo gate sequence, an optional analysis pulse with scanned phase, and state-selective shelving and readout.

The Raman beam geometry we use is summarised in Fig. 8.3 (see Fig. 5.7 for full details). The $R_\parallel$ and R_H beams are used for driving motion-coupled qubit transitions for cooling and temperature diagnostics. The gate force itself is produced by the R_V and $R_\parallel$ beams; these are polarised so as to maximise the gate force. The polarisation of the $R_\parallel$ beam is fine-tuned with a $\lambda/4$ waveplate so that it produces no 'single beam' differential light-shift on the qubit states (the R_V and R_H beams are constrained by

Fig. 8.3 Raman beam geometry. The gate is driven by the R_V and $R_\parallel$ beams, derived from a single laser, with a difference frequency $\omega = \omega_z + \delta$, where $\delta = 4\pi/t_g$ is the gate detuning for a 2-loop gate of total duration t_g. The R_H and $R_\parallel$ beams are used for sideband cooling, and have a difference frequency $\approx$3.2 GHz. The static magnetic field $\mathbf{B_0}$ defines the quantisation axis, while $\mathbf{z}$ shows the axis of the linear trap. The lattice k-vector of the two Raman beams has no projection in the radial direction, hence the Raman beams do not couple to the radial modes

the geometry to not produce any differential light-shift). The beam alignment and stability characterization is described in Sect. 6.4. We perform the gate on the axial centre-of-mass mode ($f_z \sim 1.96$ MHz) in preference to the axial breathing mode to avoid errors caused by Kerr cross-coupling (Sect. 4.4.8).

8.3.1.1 Setting the Ion Spacing

We need to adjust the ion spacing (i.e. the axial trap frequency) so the two ions see a π phase difference in the spin-dependent gate force. If the phase difference is not exactly π a small near-resonant force is applied to the states $|00\rangle, |11\rangle$ leading to a geometric phase quadratic in this force. This additional geometric phase reduces the difference in the geometric phase acquired between $|10\rangle, |01\rangle$ and $|00\rangle, |11\rangle$ requiring a larger Rabi frequency to complete the gate in a given time, in turn leading to a higher photon scattering error. As this effect is quadratic we only need the ratio of the force between the (ideally) driven and undriven states to be ~10 to keep the excess Rabi frequency (and hence error) below $\sim1\%$. We measure the force on the driven and undriven states indirectly by measuring the light-shift force on the $|00\rangle$ state on both the centre-of-mass and breathing axial modes: for a phase difference of π there should be no force on the centre-of-mass mode but a maximal force on the breathing mode. If we apply the on-resonant light-shift force to a ground-state cooled motional mode we create a coherent state with size $\alpha \sim \Omega_{force}t$. Probing the resultant coherent state with a red-sideband π-pulse allows us to measure the size of the coherent state. We adjust the axial frequency to set the force ratio to be larger than 10. For our $f_z \sim 1.96$ MHz trap and our Raman beam geometry this corresponds to an ion spacing of 12.5 standing wave periods. A mode frequency drift of 10 kHz (much larger than we observe) would only change the force ratio so as to increase the Rabi frequency required by 2%.

8.3.1.2 Measuring Readout Errors

We make the assumption that the readout errors for the two ions are identical[2]; as the ^{43}Ca$^+$ state-selective shelving relies only on simple optical pumping and is relatively insensitive to shelving beam intensity this is a good assumption. We measure the readout level from the optically pumped initial state $|\downarrow\rangle = S_{1/2}^{4,+4}$ by performing the readout process and counting the number of incorrect results. We measure the readout level for the state $|\uparrow\rangle = S_{1/2}^{3,+3}$ by optically pumping into the $F = 3$ manifold [3] using a weak ($I \sim I_0$) beam resonant with the $S_{1/2}^4 - P_{1/2}^4$ transition. This leaves

---

[2]This means we ignore errors from shelf decay during the readout. These errors lead to an $\sim10^{-4}$ under-estimate of fidelity (see Appendix C.3).

[3]We do not pump into any particular state in the $F = 3$ manifold, but rely on the insensitivity to m_f of the off-resonant shelving of $F = 3$ states.

$\sim 10^{-4}$ of the population in the $F = 4$ manifold, but this systematic effect is smaller than our typical statistical errors so we ignore it.

The 'readout errors' measured like this include all the state-detection errors, but exclude most of the state-preparation error. The main state erroneously populated by the state preparation is $S_{1/2}^{4,+3}$. The population in this state is shelved by the readout process, hence is not detected as an error.

We measure the readout errors by preparing $|\downarrow\downarrow\rangle$ and $|\uparrow\uparrow\rangle$, and reading out, normally using 4×10^4 repetitions for each state. Our typical readout errors are 8×10^{-4} from $S_{1/2}^{3,*}$ and $\approx 20 \times 10^{-4}$ from $S_{1/2}^{4,+4}$, hence we determine the readout errors with a statistical uncertainty of $\approx 2 \times 10^{-4}$. Using these readout errors, and the inverse map calculated in Appendix C, we infer the true spin-state. If we define the mean readout error $\bar{\epsilon} = \frac{1}{2}(\epsilon_\downarrow + \epsilon_\uparrow)$ and assume the readout errors are small, the infidelity of a perfect Bell state measured with imperfect readout is $1 - \mathcal{F} \approx 3\bar{\epsilon}$ (Appendix C).

8.3.1.3   Setting the Light-Shift Phase

When we have the gate duration set correctly for our two-loop gate, each half of the gate pulse (in each arm of the spin-echo) drives a closed loop in phase space, and a difference in the phase of the force between each arm does not change the gate fidelity. However if the loops are not closed (for example, when we are scanning the gate duration to optimise it) the dynamics depend significantly on the relative phase as the displacements from each gate pulse can interfere constructively or destructively. We thus need to actively control this phase.

The light-shift force we apply is off-resonant with the motional mode by detuning δ. If the time from the end of the first half of the gate pulse to the start of the second half of the gate pulse (in which we insert the spin-echo π pulse, see Fig. 8.2) is t, the relative phase of the force will have shifted by δt. In addition, the π pulse swaps the two states $|10\rangle \leftrightarrow |01\rangle$. The direction of the force is opposite on these two states, which adds an additional π phase shift to the force. We thus adjust the relative Raman phase of the two gate pulses by $\pi - \delta t$ to ensure the force phase is continuous over the two halves of the gate.

8.3.1.4   Setting the Gate Detuning and Area

To set up a gate of a given duration we first set the detuning δ and Rabi frequency Ω roughly to the calculated values. We then optimise the gate duration to ensure the motional state is separable from the spin. We do this by looking at the 'single spin-flip' signal, $P_{01} + P_{10}$: if the gate duration is correct this should be zero. For a mis-set gate duration the gate error is equal to the single spin-flip signal ($\epsilon_g = P_{01} + P_{10}$). We typically scan the gate duration by about 10% around the desired value and fit $P_{01} + P_{10}$ to determine the best duration to $\sim 0.1\%$, leading to a systematic infidelity of $\sim 10^{-4}$.

Fig. 8.4 Shaped and unshaped pulses, measured with a 1 GHz bandwidth photo-diode. The calculated gate Rabi frequencies ($\propto \sqrt{P_V}$, assuming the other Raman beam is at fixed power) for the two shaped pulses show a reasonable agreement with 0.75 μs and 1.5 μs sin^2 envelopes

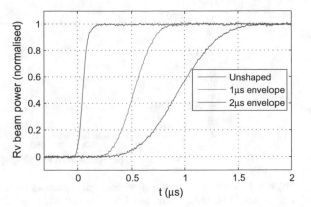

We now need to ensure the accumulated geometric phase is correct. The geometric phase scales as $\phi_{geo} \propto \Omega^2 \propto P_V P_{\|}$, where P_V, $P_{\|}$ are the Raman beam powers. We measure the geometric phase acquired using the population signals at the end of the spin-echo sequence. If there are no other sources of error $P_{00} \approx \frac{1}{2} - \alpha$, $P_{11} \approx \frac{1}{2} + \alpha$ and the gate infidelity is $\epsilon = \alpha^2$, where $\alpha = 2\Delta\phi_{geo}$. Including decoherence $P_{01} + P_{10}$ is non-zero, however the fidelity should always be maximised by setting $P_{00} = P_{11}$. To balance the populations we adjust the power of one of the Raman beams (in which the population imbalance is linear). We typically balance the populations to $\sim 1\%$ which leads to a systematic infidelity of $\sim 10^{-4}$.

8.3.1.5 Shaping the Gate Pulses

The gate Rabi frequency needs to be pulse-shaped to suppress off-resonant excitation (Sect. 4.4.11). We shape the power applied to the R_V Raman beam AOM[4] using our coherent DDS system (Appendix B.1). The $R_{\|}$ power is not shaped, and we ensure this beam turns on before and turns off after the R_V beam. Figure 8.4 shows the turn-on transient of the Raman beam power with and without pulse shaping. We shape the power applied to the AOM to give a gate Rabi frequency described by the function $\Omega/\Omega_0 = \sin^2 \frac{\pi t}{2\tau}, 0 < t < \tau$. We normally use a shape duration of $\tau = 1.5 \mu$s.

8.3.2 Single-Qubit Detuning Error

With the axial trap frequency used in these experiments the qubit frequencies of the two ions are split by $2\Delta_z = 4.906(4)$ kHz as a result of the axial magnetic field gradient (Sect. 6.1.2). Although this does not affect the gate operation itself, it does

---

[4]We use the R_V AOM as it is single pass, giving a much cleaner step response than the double-pass $R_{\|}$.

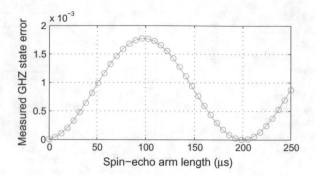

Fig. 8.5 Bell state error for a perfect phase gate inside a spin-echo with splitting between ions of $2\Delta_z = 4.906\,\text{kHz}$ and $t_\pi = 6.1\,\mu\text{s}$. The spin-echo arm length is defined as the time between the end of the first $\pi/2$ pulse and the start of the π pulse. The open circles are the numerical results and the solid line is a sinusoidal model

affect the single-qubit operations used, along with the gate, to generate the Bell state $|00\rangle + |11\rangle$. We tune the microwave local oscillator (LO) symmetrically between qubits, so that the LO is detuned by $\pm\Delta_z$ from the two qubits. The Rabi frequency of the microwaves is $\Omega/2\pi = 82\,\text{kHz}$ ($t_\pi = 6.1\,\mu\text{s}$), thus the infidelity of a single π ($\pi/2$) pulse is $\approx 10^{-3}$ (10^{-4}). With a spin-echo sequence these errors coherently add or subtract, depending on the relative phase of the pulses; if all the pulses are in phase the total error is only 10^{-6}. However as we change the length of the spin-echo sequence the relative phase of the pulses change due to the detuning of the LO, this gives rise to an error dependent on spin-echo length. To quantify this, we perform numerical simulations to determine the error in producing a Bell state from $|00\rangle$ assuming a perfect phase gate. The results of these simulations (Fig. 8.5) show that the error is well described by a sinusoidal model. These simulations include the error in the analysis pulse, but we find that the error from the imperfect analysis pulse is small compared to the imperfect spin-echo sequence. As this error is determined by the sequence timing, the qubit frequency splitting, and the microwave Rabi frequency, which are all known accurately, we can be confident in the calculation of this error. If we are interested in measuring the fidelity of the *gate* operation, we can thus subtract this single-qubit error.

We note that there are several other sources of error from the single-qubit operations, such as imperfectly set pulse areas and off-resonant excitation to other states in the ground level manifold. These $\sim 10^{-4}$ sources of error are not included in our modelling.

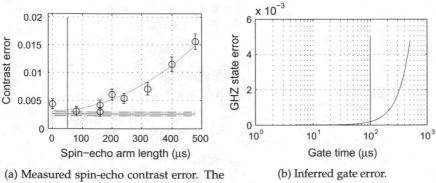

(a) Measured spin-echo contrast error. The blue lines give the readout level and uncertainty.

(b) Inferred gate error.

Fig. 8.6 Spin dephasing error. We measure the single-ion spin-echo contrast error and fit a quadratic model (**a**), and calculate how this would affect our measured Bell state error (**b**). In both plots the *red line* indicates the typical gate duration

8.3.3 Spin-Dephasing Error

We expect spin-dephasing from magnetic field noise that is uncorrelated between the two arms of the spin-echo to give errors (though it does not affect the gate operation itself). To characterize this, we measure the peak contrast of a spin-echo sequence (with no gate pulses) versus arm length on a single ion, with the first pulse of the spin-echo starting at the same delay from the line-trigger as in the gate experiments. The results are plotted in Fig. 8.6a, along with the minimum contrast error expected from the readout levels and a fit to a quadratic model.

For a single ion the reduction in spin-echo contrast from the spin-dephasing gives an identical result on any measured signal to a readout error of $\bar{\epsilon} = \frac{1}{2}(1-c)$, where c is the single-ion spin-echo contrast. As our light-shift gate commutes with the σ_z dephasing term this result holds for our gate sequence. The measured Bell state error from a readout error of $\bar{\epsilon}$ is $3\bar{\epsilon}$ (Appendix C.4), hence the dephasing error contribution to the measured Bell state fidelity is $\epsilon = \frac{3}{2}(1-c)$ (Fig. 8.6b). This model suggests that for $t_g = 500\,\mu$s the Bell state error from spin dephasing is 0.5%; slow gate experiments will be limited by this spin-dephasing. For $t_g = 100\,\mu$s the dephasing error is already below 2×10^{-4} – for gate lengths comparable to or shorter than this spin-dephasing will not be an issue.

8.3.4 Fitting Parity Scans

If we want to measure the fidelity of the experimental state to a known phase ϕ of Bell state, we can just measure the parity for two different analysis pulse phases (0

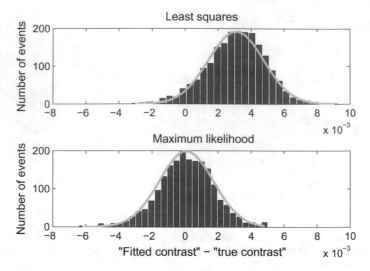

Fig. 8.7 Monte-Carlo simulation of least-squares fitting and maximum-likelihood fitting for a sinusoid $\frac{c}{2}\sin\phi$ with contrast $c = 0.995$ sampled with $N = 500$. The *blue* histograms show the fitted contrast for the 2000 synthetic datasets. The *red curve* is a Gaussian centred on the mean fitted value, with width given by the mean error bar returned by the fit function (color figure online)

and $\pi/2$). Often the offset of this phase is poorly known and we scan ϕ to map out a parity fringe, then fit this fringe to obtain the magnitude $|\rho_{11,00}|$. For a high-fidelity gate the contrast of this fringe is very near 1; here we consider how we can fit this while avoiding a systematic over- or under-estimate of the contrast. We can model the parity measurement as a Bernoulli trial giving the results 'even parity' or 'odd parity', so this has exactly the same statistics as a normal single-spin measurement. The binomial distribution is significantly non-Normal: in the following we will see that if we fit data in a naïve way using a least-squares fit (which assumes a Normal distribution) we can introduce significant biases into the data. Using a maximum likelihood method[5] instead removes this bias.

To test our data analysis routines we generate synthetic datasets consisting of N Bernoulli trials for each phase setting, sampled with probability given by a contrast parameter. We then fit these datasets and test to see how accurately the fitting methods estimate the true contrast, and how accurate the uncertainty estimates calculated by the fitting routine are. An example synthetic dataset is plotted in Fig. 8.7. This shows that, for these parameters, the least-squares method systematically overestimates the contrast by ≈0.3%, whereas the maximum-likelihood method is unbiased. This also shows that for both methods the error bars estimated by the fitting routines agree well with the actual fit uncertainty.

---

[5]In the maximum likelihood method we maximise the log-likelihood function given by $\log \mathcal{L} = \sum_i B(k_i; N, p_i)$, where B is the Binomial distribution probability distribution function, with k_i the number of successes in N trials, and p_i the model success probability.

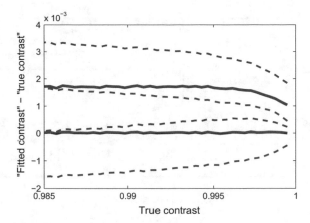

Fig. 8.8 Accuracy of the maximum-likelihood (*blue*) and least-squares (*red*) method versus sinusoid contrast sampled with $N = 1000$. The *solid lines* give the fitted contrast error and the *dashed lines* give the 1σ error bars calculated by the fitting routine. The maximum likelihood method gives an unbiased estimate of the true contrast, whereas the least-squares method overestimates the contrast by $> 1\sigma$

Figure 8.8 shows how the fitting methods compare over a range of contrasts $0.985 < c < 0.999$. For this calculation we sample the sinusoid only at 15 points spread over a sixth of a period at both the top and bottom of the fringe, as we do for high-fidelity experiments. This calculation shows that the maximum-likelihood method gives an unbiased estimate for the parameter range we are interested in, in contrast to the $> 1\sigma$ overestimate of the least-squares method.

8.3.5 Photon Scattering Errors

We expect one of the major sources of error in our gate to be photon scattering from the Raman beams. Figure 8.9 is a plot of the theoretical error and its constituents. This calculation is based on the results of Chaps. 3 and 4. Rather than simulating all of the scattering pathways between the ground states of the ion, we assume that all Raman scattering pathways out of one qubit state lead to the other (that is, the qubit is closed apart from scattering to the D states). This is a reasonable approximation as the average scattering rate back to the S manifold but out of the qubit is 26% of the total Raman scattering rate – even if all the scattering pathways led outside the qubit this would only change the error coefficient from $\frac{3}{2}$ to 2 (see Sect. 4.4.10).

We want to experimentally confirm this model. To isolate the error from photon scattering we keep the gate duration constant while varying the Raman detuning Δ and adjusting the Raman beam power P to keep the gate Rabi frequency $\Omega \propto P/\Delta$ constant. As the gate duration is kept constant any off-resonant errors, spin-echo errors, spin dephasing errors, motional errors etc. are kept constant; any change in

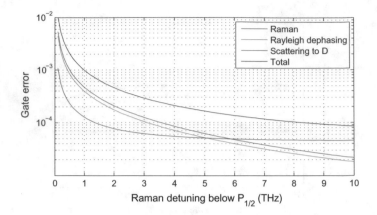

Fig. 8.9 Theoretical photon scattering error for a two-qubit light-shift gate on the 1.96 MHz centre-of-mass mode versus Raman detuning. The error from scattering to the D states is equal to the error from Raman scattering and Rayleigh dephasing at a detuning of $\Delta = -9$ THz. The asymptotic error for large detuning (from scattering to the D states) is 5×10^{-5}

gate error is from the change in photon scattering. The results of this experiment are plotted in Fig. 8.10. An offset of 2.6×10^{-3} has been added to the model to account for all of the error sources that do not vary with Raman detuning. Assuming the two ions are equally spaced about the axial micro-motion null they will have an axial micromotion amplitude of $u = 38$ nm. This gives rise to a reduction in the carrier coupling strength of $\Omega / \Omega_0 = J_0(u \, \Delta k) \approx 0.83$. This means we need to use a higher Raman Rabi frequency for a given gate duration; the theory curve has been corrected to account for this. The gate error we measure agrees very well with our theory.

8.3.6   Results

Examples of the gate dynamics we measure are given in Figs. 8.11 and 8.12. They show very good agreement with the analytic models (that have no dissipation). This means that we have good control over all the relevant experimental parameters. The dynamics scan of Fig. 8.12 took 130 min to acquire; that this data shows no systematic deviations from the model (apart from those expected from photon scattering) shows that the parameters did not drift significantly over the acquisition time.

We now want to investigate how well we understand the sources of error in the gate. Many of the sources of the error scale with the gate duration, some increasing error with increasing speed (e.g. off-resonant excitation) and some decreasing error with increasing speed (e.g. motional heating). The error from photon scattering scales as $\epsilon \propto 1/\Delta$, and for a fixed power $t_g \propto \Delta$ (assuming the Raman detuning is not comparable to or larger than the fine-structure splitting ω_f). Experimentally we fix the Raman beam powers to the maximum value achievable over a wide range of

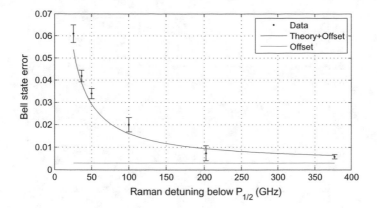

Fig. 8.10 Measurement of Bell state error from photon scattering. The gate duration is fixed at 39 μs, the Raman detuning is varied, and the Raman beam power adjusted so as to keep the Rabi frequency constant. The only change in the gate error is from the change in photon scattering error. The *blue line* is the theoretical scattering model with an offset (2.6×10^{-3}, *green line*) added to account for the remaining error sources. The theory curve includes the increase in scattering error from the axial micro-motion ($\Omega/\Omega_0 \approx 0.83$)

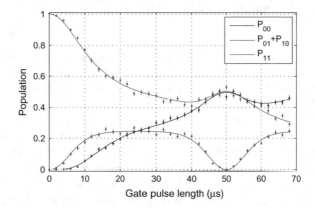

Fig. 8.11 Population dynamics for a nominally 50 μs duration two-qubit gate. The spin-echo arm length is constant and we split the gate pulse equally into both arms. The *solid line* is the analytic model of Sect. 4.2, without any dissipation terms, fitted by floating the gate detuning and Rabi frequency. At 50 μs the populations are consistent with the maximally entangled state $\frac{1}{\sqrt{2}}(|00\rangle + |11\rangle)$

Raman detunings, and adjust the Raman detuning to set the gate duration; as our desired gate duration increases we can increase the Raman detuning, reducing the scattering error, while keeping the Raman beam power constant. The results of these experiments are given in Fig. 8.13. All these data were taken with Raman beam waists of $w_0 = 27\,\mu$m and powers of $P_V = P_\parallel = 5\,$mW. The two gate halves were positioned in a spin-echo with arm-lengths a few μs longer than the gate pulse. The

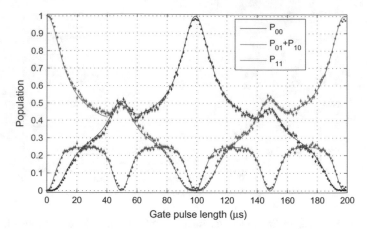

Fig. 8.12 Population dynamics as in Fig.8.11, but extending up to $4t_g$. The states at $\{0, 50\,\mu s, 100\,\mu s, 150\,\mu s, 200\,\mu s\}$ are $\{|00\rangle, \frac{1}{\sqrt{2}}(|00\rangle + |11\rangle), |11\rangle, \frac{1}{\sqrt{2}}(|00\rangle - |11\rangle), |00\rangle\}$. There is a systematic mis-set in gate Rabi frequency; after 1 gate the population imbalance is 1% (which would give a $\approx 10^{-4}$ infidelity in the absence of other errors). The deviation from the model at large times is due to photon scattering errors. There were $N = 2000$ experimental repetitions per data point, and this data took 130 min to acquire at 50 Hz repetition rate

gate duration is defined as the total length of the two gate pulses, measured from the 50% rise/fall point on the (pulse-shaped) gate Rabi frequency. The gate errors given are after normalising out the readout levels, and after subtracting the 'single-qubit detuning' error (Sect. 8.3.2).

The solid lines on Fig. 8.13 are the expected sources of error. We calculate the scattering rate error based on the previously discussed model (including the reduction of Rabi frequency from the axial micro-motion). For all the gates over 154 μs in length ($\Delta = -4$ THz) the Raman scattering error is negligible compared to dephasing errors, so we fix the Raman detuning and instead decrease the Raman beam power. For equal powers in both Raman beams we measure a coupling strength asymmetry of $g_V/g_{\parallel} = 0.86$ which leads to a negligible (1%) increase in scattering error over the balanced case. The off-resonant light-shift error is calculated for the 1.5 μs $\sin^2$ envelope used for this dataset. The heating rate for the two-ion centre-of-mass mode is taken to be 2 s^{-1} based on our measured heating rate of 1 s^{-1} (Sect. 6.5.3) for the single-ion centre-of-mass mode. We assume the motion dephasing time constant is $\tau = 200$ ms; this is pessimistic, as our measurements (Sect. 6.5.4) suggest that most of the measured dephasing is slow drift (which does not reduce gate fidelity) rather than dephasing over a single shot of the experiment.

The gate errors we measure are systematically larger than that predicted from our error budget. This is what we expect, as the error budget gives us the lowest error achievable in this apparatus, excluding many small systematic effects (for example, errors in the spin-echo sequence pulse areas). These many small effects are difficult to fix in the experiment, due to the amount of data that needs to be accumulated to determine if small adjustments have reduced the errors. The dominant source of

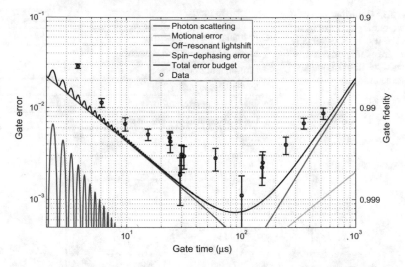

Fig. 8.13 Gate results and error budget. For $t_g \leq 154\,\mu$s, the Raman beams (5 mW, $w_0 = 27\,\mu$m) are held at constant power and the Raman detuning varied to change the gate duration (and scattering error). For $t_g > 154\,\mu$s, the Raman detuning is constant and the Raman beam power is varied. We measure the Bell state error, normalise out readout error, and subtract the 'single-qubit detuning error' to calculate the 'gate error'

error for gates slower than $200\,\mu$s is the spin-dephasing; here the deviations from the model we observe are to be expected, as our dephasing model is crude. The deviation from the model for the faster gates may be due to off-resonant excitation from imperfect pulse-shaping. The pulse-envelope time we use is modelled to give a 10^3 reduction in off-resonant excitation; moderate deviations from this pulse-shape can easily reduce the error suppression by a factor of 2.

8.3.7 Analysis of the Best Gate

The lowest gate error we measure is $\epsilon_g = 0.0011(7)$ at $t_g = 100\,\mu$s (Raman detuning $\Delta = -3.025$ THz). The error of the Bell state we produce is 0.0025(7), and the single-qubit detuning error we subtract is 0.0014 The error budget (Table 8.1) predicts a total gate error of $\epsilon_g = 0.0008$.

We estimate the laser phase and frequency noise error for a white noise floor of -100 dBc/Hz, as supported by the measurements of Sect. 6.4.2. Our experimental data is in good agreement with the predicted error considering that we have ignored most of the sources of error in the single-qubit operations (such as systematic offsets in the spin-echo pulse area, off-resonant excitation in the spin-echo pulses, etc.).

The parity scans we measure for this best gate are given in Fig. 8.14. We combine the fitted amplitude of both measurements in the calculation of the fidelity. In doing

Table 8.1 Contributions to the two-qubit gate error ϵ_g at $t_g = 100$ μs. The 'systematic errors' are estimated errors from mis-set Rabi frequency or gate detuning

Source	Error ($\times 10^{-4}$)
Raman photon scattering	4
Motional heating and dephasing	2
Motional temperature	0.4
Systematic errors	~1
Off-resonant excitation	<0.1
Laser phase and amplitude noise	0.1
Total	8

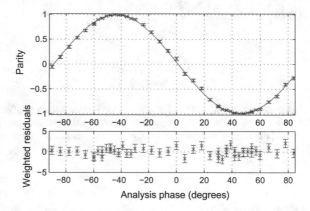

Fig. 8.14 Parity scan of the Bell state generated by a 100 μs gate. The *blue data* is a high resolution ($N = 2000$ shots per point) split scan over the peak and trough of the fringe. The red data is a lower resolution ($N = 1000$) scan over 180°. The split scan fit gives a contrast of 0.9951(13) and a phase offset of 2.2(4)°. The full scan (taken 30 min later) gives a contrast of 0.9960(23) and a phase offset of 3.8(4)°

this we ignore the difference in Bell state phase between the two measurements (1.6°), which we attribute to a small change in magnetic field between the two measurements (a qubit frequency shift of ~200 Hz would be sufficient to generate this phase shift). We justify floating the Bell state phase, rather than comparing our generated state to a fixed phase Bell state (as one might want in a computational setting), as we are primarily interested in the quality of the phase gate, rather than errors due to small phase shifts in the single-qubit gates.

8.3.8 Multiple Gates

As a further test of our error model we measure the fidelity of the Bell state produced after (an odd number of) consecutive gate operations. If we keep the spin-echo length constant (long enough to contain N gates) and then measure the resultant Bell state

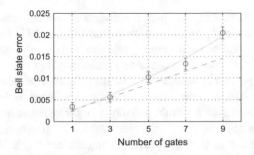

Fig. 8.15 Bell state error vs (an odd) number of two-qubit phase gates, using $t_g = 30\,\mu$s. The *dashed line* is the prediction from the error model of 15×10^{-4} per gate, while the solid line is a quadratic fit which allows for a systematic error in the Raman beam intensity of 0.5% (consistent with the precision to which it can be set); this contributes an error of 6×10^{-5} for a single gate)

fidelity for $1 \ldots N$ gate pulses the only change is due to the additional gate error. We would ideally do this for our best gate ($t_g = 100\,\mu$s), but the error from spin dephasing for a $\sim$1 ms spin-echo is much larger than the gate errors and fluctuates over time. Instead we do this for $t_g = 30\,\mu$s, performing $1 \ldots 9$ gates in a $150\,\mu$s arm-length spin-echo (Fig. 8.15). Our error budget predicts an error-per-gate of 15×10^{-4}, predominantly from photon scattering.

We first analyse the data assuming a linear error model. For $1 \ldots 5$ gates the average error is $16(3) \times 10^{-4}$ per gate, and for $1 \ldots 9$ gates the average error is $20(2) \times 10^{-4}$ per gate. These errors are independent of any fitting systematics, readout error normalisation, or single-qubit errors. As can be see in Fig. 8.15 there is a clear quadratic component to the Bell state error; this is expected from the coherent addition of the gate area error.[6] We fit a polynomial with the linear term constrained to the expected error of 15×10^{-4}, and find that a quadratic term of 6×10^{-5} describes the data well (solid line, Fig. 8.15); this corresponds to a Raman beam intensity error of 0.5%.

8.3.9 Conclusion

We have explored the compromise between speed and fidelity for a two-qubit gate, and found good agreement with our error model. The best gate has a measured error of 0.11(7)%, and we have demonstrated agreement with our error model for up to 9 gates in succession. The fastest gate we implement has a duration of $t_g = 3.8\,\mu$s and an error of 2.9(2)%, nearly an order of magnitude faster than previous trapped ion implementations. These gates are implemented on a magnetically-sensitive qubit

---

[6]When the motional loops are closed the gate operation can be described as a rotation between states $|00\rangle$ and $|11\rangle$. The rotation angle error increases linearly with number of gates, but the infidelity is quadratic on this angle error.

with a relatively short coherence time; however we have demonstrated mapping to the 146 G 'clock' qubit ($T_2^* \sim 1$ min) with an error of 2×10^{-4} [HAB+14].

The dominant limitations to improving this series of experiments are not in reducing the gate error, but in reducing the error in all the other operations. With modest increases in beam power and reductions in beam waists (as are possible with our current system) we could reduce the total predicted gate error below 3×10^{-4}. The errors in the rest of the experiment are limited to much larger values than this. Additionally, in our highest Bell state fidelities we reach statistical uncertainties of $\approx 7 \times 10^{-4}$ with 20 min of data acquisition. To reduce this uncertainty to 2×10^{-4} would require over 4 h of data. Clearly it is infeasible to go much further with these 'single gate' experiments.

8.4 Mixed Species Entangling Gates

Performing an entangling gate between non-identical ions is of interest for several different reasons. Some of the most promising architectures for scaling up a quantum computer rely on high fidelity entangling operations between ions in the same trap, and reasonable fidelity remote entanglement via photonic interconnects. The best choice of logic qubit may not coincide with the best interconnect qubit – [MK13] propose using Yb^+ as the logic qubit and Ba^+ as the interconnect qubit. One then needs a mixed-species gate to transfer the entanglement from the interconnect qubit to the logic qubit.

Even for monolithic architectures where one does not use remote entanglement a mixed-species gate allows additional flexibility. For example one may want to separate the logic ion from the 'readout and state-preparation' ion, either because there is a particularly promising logic species that cannot be readout/prepared efficiently, or because one wants to avoid any light that can resonantly scatter out of the logic qubit states.[7] With high-fidelity preparation and readout on the auxiliary species and a mixed-species gate one can state-prepare and read out the logic species without scattering any photons on the logic species.

To entangle two ions of different species using a geometric phase gate we need to produce a force that depends on the combined spin state of the two ions. The most obvious way of doing this is using different laser beams to address the different species, each producing a spin-dependent force (for example, using the Mølmer-Sørensen force or light-shift force). One could then phase-lock the laser beams (e.g. via a frequency comb) such that the force on the two species has a well defined phase. A more elegant way of doing this is to use a single set of laser beams that produce a force on both species at the same time – this is only possible for a light-shift

---

[7]In a large system there are many qubits storing quantum information while others are being read out. Scattering a single resonant photon will scramble the qubit state with high probability, but one needs to scatter $\sim 10^3$ photons (depending on collection efficiency) to detect the qubit state – this makes one very sensitive to resonant beam crosstalk.

gate (because it works irrespective of the qubit frequency). We assume that our force
is near-resonant with a motional mode that involves both ions. The light-shift force
Rabi frequency for the different crystal spin-states is then

$$\Omega_{\uparrow\uparrow} = \Omega_{\uparrow 1} + \Omega_{\uparrow 2}$$
$$\Omega_{\uparrow\downarrow} = \Omega_{\uparrow 1} + \Omega_{\downarrow 2}$$
$$\Omega_{\downarrow\uparrow} = \Omega_{\downarrow 1} + \Omega_{\uparrow 2}$$
$$\Omega_{\downarrow\downarrow} = \Omega_{\downarrow 1} + \Omega_{\downarrow 2} \tag{8.3}$$

When the motional loops are closed ($\delta t = 2\pi K$) the geometric phase accumulated is
proportional to the square of these Rabi frequencies. Making no assumptions about
the Rabi frequencies, each of the different crystal spin-states has accumulated a
different phase – this is an entangling gate, but has some difficult-to-calibrate phase
shifts. If, however, we perform one motional loop, flip both spin states (with a π pulse
on each species), and perform another motional loop we symmetrise the operation,
giving the spin-dependent geometric phase

$$\Phi_{\uparrow\uparrow} = \Phi_{\downarrow\downarrow} \propto (\Omega_{\uparrow 1} + \Omega_{\uparrow 2})^2 + (\Omega_{\downarrow 1} + \Omega_{\downarrow 2})^2$$
$$\Phi_{\uparrow\downarrow} = \Phi_{\downarrow\uparrow} \propto (\Omega_{\uparrow 1} + \Omega_{\downarrow 2})^2 + (\Omega_{\downarrow 1} + \Omega_{\uparrow 2})^2 \tag{8.4}$$

If we adjust the Rabi frequency and detuning such that $\Phi_{\uparrow\uparrow} - \Phi_{\uparrow\downarrow} = \pi/2$ we have
generated a maximally entangling gate. In the following we present an experimental
demonstration of this technique with two different isotopes of calcium, and then we
discuss how this mechanism could be used to entangle Ca^+ and Sr^+.

8.4.1 Entangling $^{40}Ca^+$ and $^{43}Ca^+$

We implement the mixed-species gate on $^{40}Ca^+$ and $^{43}Ca^+$ due to the minimal
modifications to the experiment required (Sect. 5.4 describes these in detail). The
$^{40}Ca^+$ qubit frequency is ≈ 5 MHz and the $^{43}Ca^+$ qubit frequency is ≈ 3.2 GHz. The
isotope shift on the S-P transitions of ~ 1 GHz gives a scattering rate ratio (for a beam
resonant with one species) of $\sim 10^{-3}$. This allows species-selective state-preparation
[HMS+09], but not species-selective readout with our usual methods. We implement
single-qubit manipulations on $^{43}Ca^+$ with microwaves and on $^{40}Ca^+$ with Raman
lasers. The readout errors we achieve on mixed crystals are slightly higher than on
pure crystals, due to a compromise needed to read out both species at the same time:
the 850 nm laser is needed for the $^{40}Ca^+$ EIT readout, so has the wrong polarisation
and frequency for repumping the $^{43}Ca^+$ to increase the shelving efficiency. Our
typical readout errors are {6.5, 2.3%} for $^{43}Ca^+$ and {7.9, 7.4%} for $^{40}Ca^+$.

We use the same Raman beam paths as for the $^{43}Ca^+$-$^{43}Ca^+$ gate, but we bypass the
800 MHz AOM in the Raman laser master-slave injection path so that the master and

slave Raman lasers run at the same frequency (rather than with a 3.2 GHz difference) – this allows us to drive the (5 MHz) ^{40}Ca$^+$ qubit with the Raman lasers for sideband cooling and (motionally-insensitive) single-qubit operations. The Raman detuning used is $\Delta = -1.04$ THz. Due to the hyperfine splitting of the P$_{1/2}$ level in ^{43}Ca$^+$ a pure $\sigma^\pm$ polarised beam causes a small differential light-shift. We normally null this light-shift by imbalancing the σ^+ and σ^- components, but in this experiment this would cause a light-shift on the ^{40}Ca$^+$ qubit. For our Raman detuning this light-shift causes a < 10 mrad phase shift on the ^{43}Ca$^+$ qubit for a Raman π-pulse on the ^{40}Ca$^+$ qubit – this gives a negligible error.

The force Rabi frequencies for the different qubits and spin states are

$$\Omega_0 = \frac{1}{6} g_r g_b \left[\frac{\omega_f}{\Delta(\Delta - \omega_f)} \right]$$

$$\Omega_{\uparrow 40} = \eta_{40} \Omega_0$$

$$\Omega_{\downarrow 40} = -\eta_{40} \Omega_0$$

$$\Omega_{\uparrow 43} = \frac{3}{4} \eta_{43} \Omega_0$$

$$\Omega_{\downarrow 43} = -\eta_{43} \Omega_0 \tag{8.5}$$

where we have ignored the isotope shift (~ 1 GHz $\ll \Delta \approx 1$ THz) in the joint definition of Δ. Due to the mass asymmetry of the crystal the Lamb-Dicke parameters for the two ions are slightly different [Hom13] – for these experiments we use the in-phase axial mode ($f_{\text{in-phase}} = 1.997$ MHz) giving $\eta_{43} = 0.126$ and $\eta_{40} = 0.121$. For a half-integer standing wavelength ion spacing the total spin-dependent force is

$$\Omega_{\uparrow 40, \uparrow 43} = 0.027 \, \Omega_0$$

$$\Omega_{\uparrow 40, \downarrow 43} = 0.247 \, \Omega_0$$

$$\Omega_{\downarrow 40, \uparrow 43} = -0.216 \, \Omega_0$$

$$\Omega_{\downarrow 40, \downarrow 43} = 0.005 \, \Omega_0 \tag{8.6}$$

After we symmetrise the operation, as previously described, we find $\Phi_{\uparrow\downarrow}/\Phi_{\uparrow\uparrow} = 144$: we have accumulated a large spin-dependent phase of the form we desire. As $\Phi_{\uparrow\uparrow}$ is negligible compared to $\Phi_{\uparrow\downarrow}$ we only need a marginally higher gate Rabi frequency than for a ^{43}Ca$^+$-^{43}Ca$^+$ gate, hence the photon scattering error is comparable. For the Raman detuning we use ($\Delta = -1.04$ THz) we expect a scattering error of $\sim 0.1\%$.

In our experiments we sideband cool both modes ($\bar{n} < 0.1$) using the Raman lasers addressing the ^{40}Ca$^+$ ion, state-prepare both species, perform the gate and spin-echo operation ($t_\pi = 2.5, 6.1$ μs for ^{40}Ca$^+$ and ^{43}Ca$^+$ respectively), perform tomography operations, and read out both ions. The magnetic field gradient along the axis of the trap causes the qubit frequencies to change depending on the crystal order. We tune each of the qubit local oscillators into resonance for one ordering of the crystal. We monitor for ion ordering changes with a slow ($t_\pi = 100$ μs) π pulse

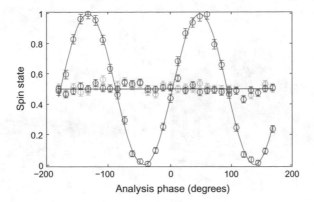

Fig. 8.16 Analysis phase scan on the Bell state produced by the ^{40}Ca$^+$-^{43}Ca$^+$ gate. The *green* and *red points* are the spin-flip probabilities for the ^{40}Ca$^+$ and ^{43}Ca$^+$ ions respectively – for a Bell state these should both be 1/2. The *blue points* are the probability of two ion spin-states being anti-aligned, and the *solid blue line* is a sinusoid fit. For a Bell state this should oscillate from 0 to 1 as $\sin 2\phi$ where ϕ is the analysis pulse phase

on one of the ions. When the ions swap order we heat the crystal up by repeatedly blue detuning the Doppler cooling beam and recrystallising until the ion order is correct.

We measure the fidelity of the Bell state produced by our gate to be $\mathcal{F} = 99.8(5)\%$ for a gate time of $t_g = 27.4\,\mu$s. The achieved fidelity (and uncertainty on the fidelity) are primarily limited by our confidence in the ^{40}Ca$^+$ readout systematics, rather than any error source intrinsic to the gate. Figure 8.16 shows the results produced by scanning the phase of the analysis pulse – this clearly shows that the state of one ion alone (tracing out the other ion's state) is maximally mixed, but that the two ions' states are strongly correlated. We also perform full state tomography of the resulting Bell state by applying independent analysis pulses to the ^{40}Ca$^+$ and ^{43}Ca$^+$ (Fig. 8.17) – this gives a fidelity of $\mathcal{F} = 99(1)\%$.

8.4.2 Future Application: Entangling ^{43}Ca$^+$ and ^{88}Sr$^+$

The two species, ^{40}Ca$^+$ and ^{43}Ca$^+$, we have used to demonstrate a mixed-species gate have transitions that are not well enough resolved to Doppler-cool or read out one species without scattering photons off the other, hence we cannot use this toy system for many of the applications in which one might want a mixed-species gate. In this section we discuss the potential to implement a gate between ^{43}Ca$^+$ and ^{88}Sr$^+$. The atomic structure of ^{88}Sr$^+$ is suited to remote entanglement via photonic interconnects due to the lack of nuclear spin. An entangling gate between ^{88}Sr$^+$ and ^{43}Ca$^+$ would allow entanglement swapping from a good interconnect ion to a good logic ion. This is hence a very interesting combination of species to pursue.

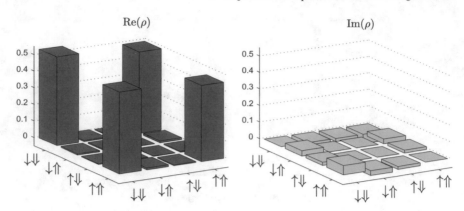

Fig. 8.17 Density matrix elements ρ of a ^{40}Ca$^+$-^{43}Ca$^+$ Bell state obtained by full tomography. $\{\Downarrow, \Uparrow\}$ is the ^{43}Ca$^+$ qubit state, and $\{\downarrow, \uparrow\}$ is the ^{40}Ca$^+$ qubit spin state. This density matrix is consistent with the Bell state $|\Uparrow\uparrow\rangle + |\Downarrow\downarrow\rangle$ to within the systematic errors from the imperfect tomography pulses

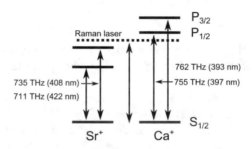

Fig. 8.18 Transitions in Sr$^+$ and Ca$^+$ relevant to a mixed-species light-shift gate. The $S_{1/2}\leftrightarrow P_{1/2}$ transition in calcium is separated from the $S_{1/2} - P_{3/2}$ transition in strontium by 20 THz. A Raman laser tuned midway between these transitions could generate a spin-dependent force on both species at the same time

The structure of the S and P levels of Ca$^+$ and Sr$^+$ is given in Fig. 8.18. The splitting between the $S_{1/2}\leftrightarrow P_{1/2}$ Doppler-cooling transitions is 44 THz – a resonant beam that cools one species does not affect the other at any appreciable level. More interesting is the splitting of 20 THz between the $S_{1/2}\leftrightarrow P_{1/2}$ transition in calcium and the $S_{1/2}\leftrightarrow P_{3/2}$ transition in strontium. A pair of Raman beams of reasonable intensity tuned between these transitions could couple to both species – the largest Raman detuning we used for the ^{43}Ca$^+$-^{43}Ca$^+$ gates was 4 THz, comparable to the 10 THz needed to couple equally to the two species.

For a ^{43}Ca$^+$-^{88}Sr$^+$ crystal with an in-phase axial mode frequency of $f = 2$ MHz the Lamb-Dicke parameters are $\eta_{Ca} = 0.078$ and $\eta_{Sr} = 0.107$. A crude estimate gives an optimum (lowest gate time for fixed intensity) Raman detuning for an entangling gate to be -8 THz from the Ca$^+$ $S_{1/2}\leftrightarrow P_{1/2}$ transition. At this detuning the scattering error is $\sim 10^{-4}$, and the Raman intensity needs to be a factor ~ 5 higher than that we used for our best ^{43}Ca$^+$-^{43}Ca$^+$ gate to maintain a gate time of 100 µs. This is easily achievable.

References

[GMT+12] Gaebler, J.P., A.M. Meier, T.R. Tan, R. Bowler, Y. Lin, D. Hanneke, J.D. Jost, J.P. Home, E. Knill, D. Leibfried, and D.J. Wineland. 2012. Randomized benchmarking of multiqubit gates. *Physical Review Letters* 108 (26): 260503.

[BIWH96] Bollinger, J.J., W.M. Itano, D.J. Wineland, and D.J. Heinzen. 1996. Optimal frequency measurements with maximally correlated states. *Physical Review A* 54 (6): R4649–R4652.

[SKK+00] Sackett, C.A., D. Kielpinski, B. King, C.E. Langer, V. Meyer, C.J. Myatt, M.A. Rowe, Q.A. Turchette, W.M. Itano, D.J. Wineland, and C. Monroe. 2000. Experimental entanglement of four particles. *Nature* 404 (6775): 256–9.

[Mon11] T. Monz. 2011. Quantum information processing beyond ten ion-qubits. Ph.D thesis, University of Innsbruck.

[HML+06] Home, J.P., M.J. McDonnell, D.M. Lucas, G. Imreh, B.C. Keitch, D.J. Szwer, N.R. Thomas, S.C. Webster, D.N. Stacey, and A.M. Steane. 2006. Deterministic entanglement and tomography of ion-spin qubits. *New Journal of Physics* 8 (9): 188–188.

[MSW+04] McDonnell, M.J., J.-P. Stacey, S.C. Webster, J.P. Home, A. Ramos, D.M. Lucas, D.N. Stacey, and A.M. Steane. 2004. High-efficiency detection of a single quantum of angular momentum by suppression of optical pumping. *Physical Review Letters* 93 (15): 1–4.

[HAB+14] Harty, T.P., D.T.C. Allcock, C.J. Ballance, L. Guidoni, H.A. Janacek, N.M. Linke, D.N. Stacey, and D.M. Lucas. 2014. High-fidelity preparation, gates, memory, and readout of a trapped-ion quantum bit. *Physical Review Letters* 113 (22): 220501.

[MK13] Monroe, C., and J. Kim. 2013. Scaling the ion trap quantum processor. *Science* 339 (6124): 1164–1169.

[HMS+09] Home, J.P., M.J. McDonnell, D.J. Szwer, B. Keitch, D.M. Lucas, D. Stacey, and A.M. Steane. 2009. Memory coherence of a sympathetically cooled trapped-ion qubit. *Physical Review A* 79 (5): 050305.

[Hom13] Home, J.P. 2013. Quantum science and metrology with mixed-species ion chains. *Advances In Atomic, Molecular, and Optical Physics* 62: 231–277.

Chapter 9
Conclusion

9.1 Summary

In this thesis we have described work implementing high-fidelity laser-driven two-qubit and single-qubit gates in ^{43}Ca$^+$ hyperfine qubits. The best single-qubit and two-qubit gate errors we achieve are at least an order of magnitude below the fault-tolerant threshold for the best surface code error correction algorithms.

The single-qubit gates are predominantly limited by the Raman laser differential phase noise. This noise could easily be removed with the feedback loop described in Sect. 6.4.2. The next leading source of error is photon scattering; using similar Raman detunings to that we used for our two-qubit gates this could be reduced below an average error-per-gate of 10^{-5}.

The two-qubit gates are nominally limited by photon scattering and motional dephasing; by increasing beam intensities and the Raman detuning Δ the sum of these could be reduced below 4×10^{-4}. However in our current experiment this is difficult to measure due to the systematic errors at the 10^{-3} level. To avoid these errors we intend to perform two-qubit randomized benchmarking [GMT+12]: this will allow us to amplify the gate errors, and also greatly decrease the time taken to measure the gate fidelity. This will enable real-time optimisation of the gate parameters, and should enable us to measure errors below the 10^{-3} level.

In this thesis we have also experimentally demonstrated a mixed-species entangling gate between ^{40}Ca$^+$ and ^{43}Ca$^+$. Apart from the inherent novelty of entangling non-identical particles, we anticipate that this gate mechanism, applied for example to ^{43}Ca$^+$ and ^{88}Sr$^+$, will be a useful building block for a scalable quantum computer.

9.2 Comparison with Other Results

In this section we compare our single-qubit gate and two-qubit gate results with selected other results, both in trapped-ion systems and other implementations.

© Springer International Publishing AG 2017
C.J. Ballance, *High-Fidelity Quantum Logic in Ca$^+$*,
Springer Theses, DOI 10.1007/978-3-319-68216-7_9

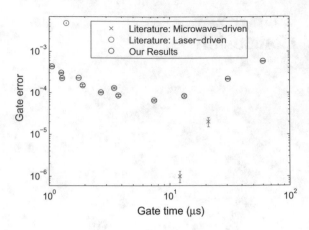

Fig. 9.1 The best single-qubit gate results from the literature, alongside our results. All of these errors were measured by randomised benchmarking. The 'gate time' is the length of the $\pi/2$ pulses used to implement the randomised Clifford gates

9.2.1 Single-Qubit Gates

The lowest error single-qubit gates implemented in trapped-ion systems are plotted in Fig. 9.1, alongside our results.

The lowest single-qubit error gate achieved in a trapped-ion system is $1.0(3) \times 10^{-6}$ [HAB+14]. This was implemented by my colleagues and me in a ^{43}Ca$^+$ hyperfine qubit using microwaves. A separate experiment achieved an error of $2.0(2) \times 10^{-5}$ in a ^{9}Be$^+$ hyperfine qubit using microwaves [BWC+11]. These errors are significantly lower than the $6.6(3) \times 10^{-5}$ error we measure in this thesis using laser-driven gates. However there are difficulties in using microwaves for 'addressed' single-qubit gates (implementing a gate on one qubit while not affecting neighbouring qubits), due to the microwave wavelength (~ 10 cm) being far larger than the inter-ion separation (though there are schemes to work past this [WOC+13, ALH+13, PSVW14]).

Laser beams, however, can be focused down below the inter-ion separation, allowing 'direct' addressing. The previous lowest error laser-driven single-qubit gate is $4.8(2) \times 10^{-3}$ [KLR+08] – an error nearly two orders of magnitude larger than our result.

Low error-rates have been realised in two other implementations. Using superconducting Josephson junctions, an addressed single-qubit gate error of 6×10^{-4} has been achieved in an array of 5 qubits [BKM+14], and in an ensemble of ^{87}Rb atoms in an optical lattice, an average single-qubit gate error $1.4(1) \times 10^{-4}$ has been achieved [OCNP10].

9.2.2 Two-Qubit Gates

A summary of the best trapped-ion two-qubit gates in the literature is given in Table 9.1. This is not an exhaustive listing, but it includes the best results on all

Table 9.1 Summary of the best two-qubit gate fidelities achieved in trapped-ion qubits. Where sufficient information is provided in the paper the state-preparation and readout error has been subtracted – this is marked by an asterisk. Fidelities measured by randomised benchmarking are marked with a dagger, all others are Bell state fidelities measured by partial state tomography

Qubit	Fidelity	Gate time (μs)	Reference
^{9}Be$^+$ ground level	97(2)%	39	[LDM+03]
^{40}Ca$^+$ optical	99.3(1)%	50	[BKRB08]
^{40}Ca$^+$ optical	97.1(2)%	25	[KBZ+09]
^{171}Yb$^+$ ground level clock	96(2)%	38	[KCI+09]
^{9}Be$^+$ ground level clock	93.1(17)% †	20	[GMT+12]
^{9}Be$^+$ ground level clock	98.3(4)%*	250	[TGB+13]
^{88}Sr$^+$ optical	98.5(10)%	130	[NAK+14]
^{43}Ca$^+$ ground level	99.9(1)%*	100	This work

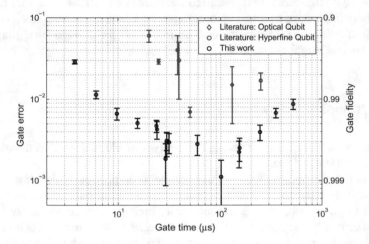

Fig. 9.2 Plot of the two-qubit gate results from the literature (listed in Table 9.1) and our results versus gate duration. Note that our results have readout levels normalised out, whereas most of the literature results do not

of the different flavours of qubits and gate mechanisms that are currently used. All of the two-qubit gates listed here are driven by lasers (as opposed to microwaves which have achieved a maximum fidelity of 90%). We also plot these gate fidelities versus the gate duration, alongside our results (Fig. 9.2).

Finally, we plot the history of two-qubit gate error in both trapped-ion systems and superconducting Josephson-junction systems (Fig. 9.3) – these are the only two systems with high-fidelity two-qubit gates along with single-shot single-qubit readout.[1] The lowest two-qubit gate error demonstrated on superconducting qubits is

---

[1] Two-qubit gate errors below 10^{-2} have been demonstrated in NMR systems, however these systems do not permit single-qubit readout, hence error-correction cannot be implemented.

Fig. 9.3 History of two-qubit gate errors in trapped-ion systems (blue crosses) and superconducting Josephson-junction systems (red circles). The result of this work is boxed (color figure online)

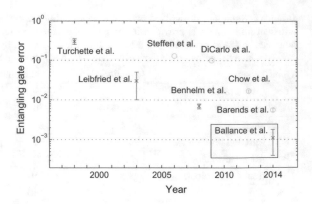

$5.6(5) \times 10^{-3}$, measured by randomized benchmarking [BKM+14]. This result is particularly impressive as it was demonstrated on a 5-qubit system (albeit only the best pair of qubits could be entangled with this error; the average entangling error over all nearest neighbour pairs was 7.4×10^{-3}).

9.3 Outlook

How far are we from implementing the freely-scalable architecture described in Sect. 1.2 with trapped ions?

One of the principal components that had not been previously demonstrated is a low error (below threshold) two-qubit gate in a long-lived qubit (the previous best was in an optical qubit with a fundamentally-limited spontaneous decay time of ≈ 1 s) – our results change this. However the magnetically-sensitive qubit in which we implement our entangling gates is not an ideal memory qubit. Although it does not have any spontaneous decay, or a fundamental coherence time limit, magnetic field noise will always be a technical limitation.

A much better memory qubit is the magnetic-field-insensitive 146 G clock qubit in $^{43}Ca^+$: in this qubit we have demonstrated coherence times of 50 s (without any magnetic shielding), and a combined state-preparation and readout fidelity of 99.93% [HAB+14]. Although we cannot directly apply our light-shift gate to this qubit, we have demonstrated mapping between the stretch 'gate' qubit and this clock 'memory' qubit with errors $\approx 2 \times 10^{-4}$. We thus expect that we could map two ions out of the clock qubit, perform an entangling gate, and map back into the clock qubit with an error below 10^{-3} with our current experimental techniques.

The work in this thesis was performed in a macroscopic ion-trap that is poorly suited for scaling up to a large array of traps, because of the mechanical size of the trap structure and the difficulty of splitting and combining multiple crystals of ions. Micro-fabricated 3D traps [BOV+09] and surface-electrode ion-traps [SCR+06] are more suitable for scaling as they can easily have an arbitrary number of trap zones, are potentially more repeatable to fabricate, and are better suited for integrated optics. These advantages come with a significant problem; the 'anomalous heating' of the ions' motion increases dramatically in 'small' traps [TMK+00], leading to increased gate errors. However, this problem can be mitigated. Heating rates have been reduced by cleaning with lasers [AGH+11] or argon ion beams [HCW+12], and by cryogenically cooling the trap electrodes [CS14]. It thus seems feasible that in the future micro-fabricated traps could be made that reliably had heating rates below $10\,s^{-1}$, allowing gate errors of order 10^{-4} for sub-100 μs two-qubit gates.

In order to implement general operations on more than a few qubits, we need to be able to shuttle ions between trap zones [RBKD+02], and to sympathetically cool the ions back to their ground state of motion [KKM+00] – these techniques have both been demonstrated in a computational context [HHJ+09]. A reasonable sympathetic coolant for $^{43}Ca^+$ is $^{88}Sr^+$: it has a comparable mass to $^{43}Ca^+$, so the motional modes are well mixed, and the laser wavelengths required for Doppler cooling and sideband cooling are all available from diode lasers.

One of the disadvantages of trapped-ion qubits is that several laser beam-paths are required for each trap. The obvious solution to this is to integrate fibre-coupled micro-fabricated optics into the trap structure itself, taking advantage of the techniques developed by the telecommunications industry. Laser beam delivery [KHC11], and ion florescence collection [BEM+11] have both been demonstrated in micro-fabricated ion traps, albeit with low efficiency, and no polarisation control.

The final ingredient for a freely scalable architecture is a way of entangling remote qubits. One way of doing this is using photons – despite the inevitable loss, heralding schemes allow one to prepare non-deterministically pairs of remotely entangled qubits [DBMM04]. This has been demonstrated with a success rate of 4.5 Hz and a fidelity of 78(3)% for two $^{171}Yb^+$ ground-level qubits [HIV+14]. The low entanglement generation rate in this experiment could be increased by increasing the photon collection efficiency with integrated optics, particularly if an optical cavity could be integrated. If the rate is still too low, we can attempt to entangle multiple pairs of ions at the same time – once we have a heralded success we can shuttle the entangled ions into the computation zone.

In this photonic interconnect scheme we need a lot of intense optical pulses to efficiently excite the spontaneous photon emission used to entangle the remote qubits. If any of this light scatters into the computation or memory zones it will corrupt the computational qubits. A way to avoid this is to use different species for the photonic interconnect ion and the logic ion [MK13]. We could also use $^{88}Sr^+$ for this purpose – this species has no nuclear spin, allowing a fast cycle time for the remote entanglement attempts, and the entanglement can easily be mapped onto a logic ion using a light-shift gate, as described in Sect. 8.4.2.

In conclusion, small arrays of trapped ions form ideal candidates for implementing a freely-scalable quantum computer, with all of the required building blocks demonstrated at the required levels of precision. A promising combination of ions to use for such a system is $^{43}Ca^+$ and $^{88}Sr^+$.

References

[GMT+12] Gaebler, J.P., A.M. Meier, T.R. Tan, R. Bowler, Y. Lin, D. Hanneke, J.D. Jost, J.P. Home, E. Knill, D. Leibfried, and D.J. Wineland. 2012. Randomized benchmarking of multiqubit gates. *Physical Review Letters* 108 (26): 260503.

[HAB+14] Harty, T.P., D.T.C. Allcock, C.J. Ballance, L. Guidoni, H.A. Janacek, N.M. Linke, D.N. Stacey, and D.M. Lucas. 2014. High-fidelity preparation, gates, memory, and readout of a trapped-ion quantum bit. *Physical Review Letters* 113 (22): 220501.

[BWC+11] Brown, K.R., A.C. Wilson, Y. Colombe, C. Ospelkaus, A.M. Meier, E. Knill, D. Leibfried, and D.J. Wineland. 2011. Single-qubit-gate error below 10 ^-4 in a trapped ion. *ArXiv* 8 (2): 5.

[WOC+13] Warring, U., C. Ospelkaus, Y. Colombe, R. Jördens, D. Leibfried, and D.J. Wineland. 2013. Individual-ion addressing with microwave field gradients. *Physical Review Letters* 110 (17): 173002.

[ALH+13] Aude Craik, D.P.L., N.M. Linke, T.P. Harty, C.J. Ballance, D.M. Lucas, A.M. Steane, D.T.C. Allcock, and D.T.C. Allcock. 2013. Microwave control electrodes for scalable, parallel, single-qubit operations in a surface-electrode ion trap. *Applied Physics B* 114 (1–2): 3–10.

[PSVW14] Piltz, C., T. Sriarunothai, a. F. Varón, and C. Wunderlich. 2014. A trapped-ion-based quantum byte with 10(−5) next-neighbour cross-talk. *Nature communications* 5: 4679.

[KLR+08] Knill, E., D. Leibfried, R. Reichle, J.W. Britton, R.B. Blakestad, J.D. Jost, C.E. Langer, R. Ozeri, S. Seidelin, and D.J. Wineland. 2008. Randomized benchmarking of quantum gates. *Physical Review A* 77 (1): 1–7.

[BKM+14] Barends, R., J. Kelly, A. Megrant, A. Veitia, D. Sank, E. Jeffrey, T.C. White, J. Mutus, A.G. Fowler, B. Campbell, Y. Chen, Z. Chen, B. Chiaro, A. Dunsworth, C. Neill, P. O'Malley, P. Roushan, A. Vainsencher, J. Wenner, A.N. Korotkov, A.N. Cleland, and J.M. Martinis. 2014. Superconducting quantum circuits at the surface code threshold for fault tolerance. *Nature* 508 (7497): 500–3.

[OCNP10] Olmschenk, S., R. Chicireanu, K.D. Nelson, and J.V. Porto. 2010. Randomized benchmarking of atomic qubits in an optical lattice. *New Journal of Physics* 12 (11): 113007.

[LDM+03] Leibfried, D., B. DeMarco, V. Meyer, D.M. Lucas, M.D. Barrett, J.W. Britton, W.M. Itano, B. Jelenković, C.E. Langer, T. Rosenband, and D.J. Wineland. 2003. Experimental demonstration of a robust, high-fidelity geometric two ion-qubit phase gate. *Nature* 422 (6930): 412–5.

[BKRB08] Benhelm, J., G. Kirchmair, C.F. Roos, and R. Blatt. 2008. Towards fault-tolerant quantum computing with trapped ions. *Nature Physics* 4 (6): 463–466.

[KBZ+09] Kirchmair, G., J. Benhelm, F. Zähringer, R. Gerritsma, C.F. Roos, and R. Blatt. 2009. Deterministic entanglement of ions in thermal states of motion. *New Journal of Physics* 11 (2): 023002.

[KCI+09] Kim, K., M.-S. Chang, R. Islam, S. Korenblit, L.-M. Duan, and C. Monroe. 2009. Entanglement and tunable spin-spin couplings between trapped ions using multiple transverse modes. *Physical Review Letters* 103 (12): 120502.

[TGB+13] Tan, T.R., J.P. Gaebler, R. Bowler, Y. Lin, J.D. Jost, D. Leibfried, and D.J. Wineland. 2013. Demonstration of a dressed-state phase gate for trapped ions. *Physical Review Letters* 110 (26): 263002.

[NAK+14] Navon, N., N. Akerman, S. Kotler, Y. Glickman, and R. Ozeri. 2014. Quantum process tomography of a Mølmer-Sørensen interaction. *Physical Review A* 90 (1): 010103.

[BOV+09] Blakestad, R.B., C. Ospelkaus, A.P. VanDevender, J.M. Amini, J.W. Britton, D. Leibfried, and D.J. Wineland. 2009. High-fidelity transport of trapped-ion qubits through an x-junction trap array. *Physical Review Letters* 102 (15): 153002.

[SCR+06]   Seidelin, S., J. Chiaverini, R. Reichle, J.J. Bollinger, D. Leibfried, J.W. Britton, J.H. Wesenberg, R.B. Blakestad, R. Epstein, D.B. Hume, W.M. Itano, J.D. Jost, C.E. Langer, R. Ozeri, N. Shiga, and D.J. Wineland. 2006. Microfabricated surface-electrode ion trap for scalable quantum information processing. *Physical Review Letters* 96 (25): 253003.

[TMK+00]   Turchette, Q.A., C.J. Myatt, B.E. King, C.A. Sackett, D. Kielpinski, W.M. Itano, C. Monroe, and D.J. Wineland. 2000. Decoherence and decay of motional quantum states of a trapped atom coupled to engineered reservoirs. *Physical Review A* 62 (5): 1–22.

[AGH+11]   Allcock, D.T.C., L. Guidoni, T.P. Harty, C.J. Ballance, M. Blain, A.M. Steane, and D.M. Lucas. 2011. Reduction of heating rate in a microfabricated ion trap by pulsed-laser cleaning. *New Journal of Physics* 13 (12): 123023.

[HCW+12]   Hite, D.A., Y. Colombe, A.C. Wilson, K.R. Brown, U. Warring, R. Jördens, J.D. Jost, K.S. McKay, D. Pappas, D. Leibfried, and D.J. Wineland. 2012. 100-Fold reduction of electric-field noise in an ion trap cleaned with in situ argon-ion-beam bombardment. *Physical Review Letters* 109 (10): 103001.

[CS14]   Chiaverini, J., and J.M. Sage. 2014. Insensitivity of the rate of ion motional heating to trap-electrode material over a large temperature range. *Physical Review A* 89 (1): 012318.

[RBKD+02]   Rowe, M., A. Ben-Kish, B. DeMarco, D. Leibfried, V. Meyer, J.a. Beall, J.W. Britton, J. Hughes, W.M. Itano, B. Jelenković, C.E. Langer, T. Rosenband, and D.J. Wineland. 2002. Transport of quantum states and separation of ions in a dual rf ion trap. *Quantum Information and Computation* 2(7): 257–271.

[KKM+00]   Kielpinski, D., B.E. King, C.J. Myatt, C.A. Sackett, Q.A. Turchette, W.M. Itano, C. Monroe, D.J. Wineland, and W. Zurek. 2000. Sympathetic cooling of trapped ions for quantum logic. *Physical Review A* 61 (3): 1–8.

[HHJ+09]   Home, J.P., D. Hanneke, J.D. Jost, J.M. Amini, D. Leibfried, and D.J. Wineland. 2009. Complete methods set for scalable ion trap quantum information processing. *Science* 325 (5945): 1227–30.

[KHC11]   Kim, T.H., P.F. Herskind, and I.L. Chuang. 2011. Surface-electrode ion trap with integrated light source. *Applied Physics Letters* 98 (21): 214103.

[BEM+11]   Brady, G.R., A.R. Ellis, D.L. Moehring, D. Stick, C. Highstrete, K.M. Fortier, M.G. Blain, R.A. Haltli, A.A. Cruz-Cabrera, R.D. Briggs, J.R. Wendt, T.R. Carter, S. Samora, and S.A. Kemme. 2011. Integration of fluorescence collection optics with a microfabricated surface electrode ion trap. *Applied Physics B* 103 (4): 801–808.

[DBMM04]   Duan, L.-M., B.B. Blinov, D.L. Moehring, and C. Monroe. 2004. Scalable trapped ion quantum computing with a probabilistic ion-photon mapping. *Quantum Information and Computation* 4 (3): 165–173.

[HIV+14]   Hucul, D., I.V. Inlek, G. Vittorini, C. Crocker, S. Debnath, S.M. Clark, and C. Monroe. 2014. Modular entanglement of atomic qubits using photons and phonons. *Nature Physics*.

[MK13]   Monroe, C., and J. Kim. 2013. Scaling the ion trap quantum processor. *Science* 339 (6124): 1164–1169.

Appendix A
Atomic Structure of ^{43}Ca$^+$

A.1 Atomic Constants

(See Table A.1).

A.2 Dipole Matrix Elements

To calculate the Raman scattering rates of Chap. 3, we need to calculate the S$\leftrightarrow$P transition matrix elements. These matrix elements, in terms of the 'stretch' matrix element μ (to be defined shortly), are

$$\langle F'm' | \hat{\epsilon}_q \cdot \mathbf{d} | Fm \rangle / \mu = (-1)^{1+2F'+J'+J+L+S+I+m} \sqrt{2F'+1} \sqrt{2F+1} \sqrt{2L+1}$$

$$\sqrt{3} \sqrt{2J'+1} \sqrt{2J+1} \begin{pmatrix} F' & 1 & F \\ m' & -q & -m \end{pmatrix} \begin{Bmatrix} J & J' & 1 \\ F' & F & I \end{Bmatrix} \begin{Bmatrix} L & L' & 1 \\ J' & J & S \end{Bmatrix} \quad \text{(A.1)}$$

Table A.1 Properties of the low-lying dipole-allowed transitions in Ca$^+$. The transition frequencies given are for ^{40}Ca$^+$. The isotope shift, $\nu_{43} - \nu_{40}$, is the difference between the tabulated frequency and the line centre in ^{43}Ca$^+$. (From [Szw09].)

Transition	Frequency (THz)	$\nu_{43} - \nu_{40}$ (MHz)	A coeff. (10^6 s^{-1})	I_{sat} (W m^{-2})
S$_{1/2} \leftrightarrow$ P$_{1/2}$	755.2227662(17)	688(17)	132	933.82
S$_{1/2} \leftrightarrow$ P$_{3/2}$	761.9050127(5)	692(19)	135.0(4)	987.58
D$_{3/2} \leftrightarrow$ P$_{1/2}$	346.000	−3464.3(3.0)	8.4	89.798
D$_{3/2} \leftrightarrow$ P$_{3/2}$	352.682	−3462.4(2.6)	0.955(6)	97.954
D$_{5/2} \leftrightarrow$ P$_{3/2}$	350.863	−3465.4(3.7)	8.48(4)	96.446

© Springer International Publishing AG 2017
C.J. Ballance, *High-Fidelity Quantum Logic in Ca$^+$*,
Springer Theses, DOI 10.1007/978-3-319-68216-7

where $\left\{ \begin{smallmatrix} \cdots \\ \cdots \end{smallmatrix} \right\}$ is the Wigner 6-J function and $\left(\begin{smallmatrix} \cdots \\ \cdots \end{smallmatrix} \right)$ is the Wigner 3-J function, S is the total electron spin (1/2 for Ca$^+$), I is the nuclear spin (7/2 for ^{43}Ca$^+$), and $\hat{\epsilon}_q$ is an element of the spherical basis ($q = m' - m$).

The 'stretch' transition matrix element is defined

$$\mu := \langle F' = I + 3/2, m' = I + 3/2 |\, \hat{\epsilon}_1 \cdot \boldsymbol{d}\, | F = I + 1/2, m = I + 1/2 \rangle \quad \text{(A.2)}$$

This is the largest matrix element, as the excited state $|F = I + 3/2, m = I + 3/2\rangle$ has only one dipole-allowed decay route.

For a two-level atom the spontaneous decay rate from the excited state is [MvdS99]

$$\Gamma = \frac{\omega_0^3 |\langle e|\, d\, |g\rangle|^2}{3\pi\epsilon_0 \hbar c^3} \quad \text{(A.3)}$$

Using this, and the $S_{1/2} \leftrightarrow P_{1/2}$ Einstein A coefficient, we can calculate the 'stretch' matrix element. For Ca$^+$ $\mu = 2.015\ e\, a_0$.

A.3 Raman Transition and Microwave Matrix Elements

(See Table A.2 and Fig. A.1).

Table A.2 Matrix elements for Raman and microwave transitions in one half of the ground-level manifold. The matrix elements for the remaining half of the manifold can be found by symmetry. The Raman matrix elements are evaluated in the limit $|\Delta| \gg \omega_{\text{HF}}$

Transition	Raman ($\times \frac{g_r g_b}{6} \frac{\omega_f}{\Delta(\Delta - \omega_f)}$)	Microwave ($\times \mu_B$)
$4, 4 \to 3, 3$	$-\frac{\sqrt{7}}{2}(b_- r_\pi + b_\pi r_+)$	$-\frac{\sqrt{7}}{2} B_-$
$4, 3 \to 3, 3$	$\frac{\sqrt{7}}{4}(b_- r_- - b_+ r_+)$	$\frac{\sqrt{7}}{4} B_\pi$
$4, 3 \to 3, 2$	$\frac{\sqrt{21}}{4}(b_+ r_\pi + b_\pi r_+)$	$-\frac{\sqrt{21}}{4} B_-$
$4, 2 \to 3, 3$	$\frac{1}{4}(b_+ r_\pi + b_\pi r_+)$	$-\frac{1}{4} B_+$
$4, 2 \to 3, 2$	$\frac{\sqrt{3}}{2}(b_- r_- - b_+ r_+)$	$\frac{\sqrt{3}}{2} B_\pi$
$4, 2 \to 3, 1$	$-\frac{\sqrt{15}}{4}(b_- r_\pi + b_\pi r_+)$	$-\frac{\sqrt{15}}{4} B_-$
$4, 1 \to 3, 2$	$\frac{\sqrt{3}}{4}(b_+ r_\pi + b_\pi r_-)$	$-\frac{\sqrt{3}}{4} B_+$
$4, 1 \to 3, 1$	$\frac{\sqrt{15}}{4}(b_- r_- - b_+ r_+)$	$\frac{\sqrt{15}}{4} B_\pi$
$4, 1 \to 3, 0$	$-\frac{\sqrt{5}}{2\sqrt{2}}(b_- r_\pi + b_\pi r_+)$	$-\frac{\sqrt{5}}{2\sqrt{2}} B_-$
$4, 0 \to 3, 1$	$\frac{\sqrt{3}}{2\sqrt{2}}(b_+ r_\pi + b_\pi r_-)$	$-\frac{\sqrt{3}}{2\sqrt{2}} B_+$
$4, 0 \to 3, 0$	$(b_- r_- - b_+ r_+)$	B_π

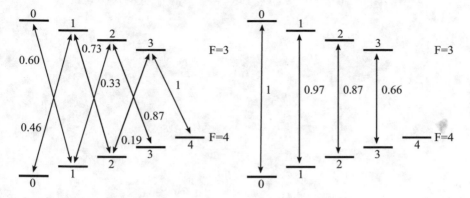

Fig. A.1 Normalised transition strengths in the ground-level manifold. These results hold for both Raman and microwave driven transitions

Appendix B
Electronics

B.1 Coherent DDS

The 'Coherent DDS' is a combination of firmware and hardware that implements RF pulse shaping and phase coherent switching over 4 channels of RF output. The hardware is a 'Milldown' DDS card,[1] which consists of an FPGA[2] connected to 4 DDS chips.[3] Each of the DDS RF outputs has a variable-gain amplifier[4] controlled by a 200 MHz update rate DAC.[5]

Each of the DDS channels has 8 frequency/phase/amplitude profiles, and an associated pulse-shape. These parameters are programmed into the card via a serial link. The active profiles for each channel, and the pulse-shaping, are controlled by a bank of TTL inputs.

B.1.1 DDS Operation and Phase Coherence

Frequency agile sources typically operate in either a phase continuous or phase coherent mode. In a phase continuous frequency switch the source simply starts accumulating phase at a rate governed by the new frequency. There is no discontinuity at the switch, and the phase of new output relative to a reference running at the same frequency depends on the time of the switch. In a phase coherent frequency switch there is a discontinuity in phase, and the new output has a well defined phase with respect to a reference oscillator – this case is equivalent to selecting between two

---

[1] Manufactured by Enterpoint.

[2] Xilinx Spartan 6 XC6SLX150T-2 FGG900.

[3] Analog Devices AD9910.

[4] Analog Devices ADL5330.

[5] Texas Instruments DAC5672A.

© Springer International Publishing AG 2017
C.J. Ballance, *High-Fidelity Quantum Logic in Ca+*,
Springer Theses, DOI 10.1007/978-3-319-68216-7

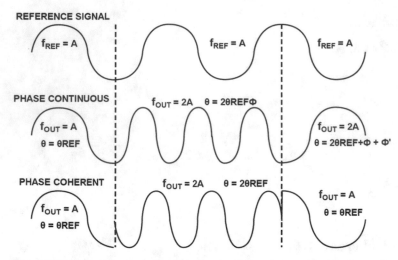

WHERE θ = PHASE OF OUTPUT SIGNAL, Φ = PHASE AT TIME OF FIRST FREQUENCY TRANSITION, AND Φ' = PHASE AT TIME OF SECOND FREQUENCY TRANSITION.

Fig. B.1 Difference between a phase continuous and phase coherent frequency switch. Adapted from AD9858 datasheet

frequency sources running continuously at different frequencies. These two modes of operation are illustrated in Fig. B.1.

In general we wish to have phase coherent sources – if we generate a pulse around the +X axis, switch to a different frequency for a while, and come back to the original frequency we wish to still be in the +X phase frame (rather than a random frame). Unfortunately DDS sources are almost universally phase continuous – it is much easier to have a single phase accumulator and merely change the rate of phase accumulation as different frequencies are selected than to have a phase accumulator for each of the different output frequencies.

B.1.1.1   AD9910 DDS Core Operation

The DDS core for the AD9910 is shown in Fig. B.2. Each of the 8 profiles has an amplitude, phase, and frequency word associated with it. Changing to a different profile changes the phase offset and frequency word, but the absolute phase of the output depends on when the profile is changed. Hence swapping between profiles programmed with the same frequency is phase coherent, but as soon as one switches to a profile with a different frequency phase coherence is lost.

The equations for the output phase of the DDS are:

$$\theta_n[31:0] = \theta_{n-1} + f[31:0] \tag{B.1}$$

$$\phi_n[18:0] = \theta_n[31:11] + \{\phi_{\text{offset}}[15:0], 0[15:0]\} \tag{B.2}$$

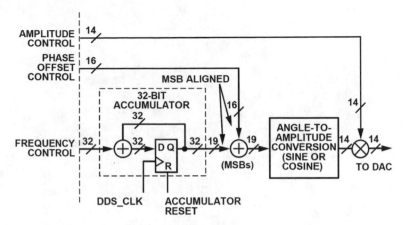

Fig. B.2 AD9910 DDS core block diagram (Adapted from AD9910 datasheet.)

where θ is the accumulator value (32 bits), f is the programmed frequency word (32 bits), ϕ_{offset} is the programmed phase word (16 bits), and ϕ is the DDS output phase (19 bits). At each time-step (one period of the 1 GHz DDS clock) a new output phase is calculated, and a new output voltage generated.

B.1.1.2 Faking Phase Coherence

We make our DDS system phase coherent by running a phase accumulator in our FPGA for each of the 8 profiles, and on a profile change reprogramming the DDS with the current frequency and phase calculated by the FPGA. One potential problem with this scheme is that the DDS phase accumulator is 32 bits, the phase output is 19 bits, but the profile phase offset (the only programmable phase) is 16 bits. This means that although the DDS and FPGA will agree on the phase exactly (as the FPGA can have an identical 32 bit phase accumulator to the DDS) the FPGA can only program the DDS with the 16 most significant bits of the phase. As the phase accumulated in the FPGA is not truncated this phase error does not build over time, it merely means that the DDS output phase will be in disagreement with the ideal signal phase by at most the equivalent to 1 bit of the 16 bit phase word. Hence $|\delta\phi| < 2\pi/2^{16} = 1 \times 10^{-4}$ rad. This is negligible in our experiments – the worst case error from this phase error for a single carrier π-pulse is $\epsilon = 10^{-8}$.

B.1.2 Pulse-Shaping

The RF output amplitude DAC (the VGA DAC) is controlled by the FPGA. On the rising and falling edges of the pulse-shape TTL signal the RF output is shaped with

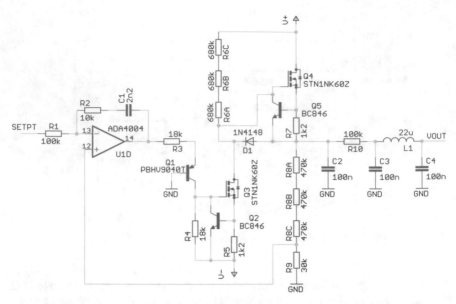

Fig. B.3 Trap DC DAC output stage. The SETPT input is a $-5\,V \to +5\,V$ signal which programs an output voltage from $-240\,V \to +240\,V$. The supplies are $V_\pm = \sim \pm 260\,V$ (Fig. B.4)

a rising and falling pulse-shape respectively. The pulse-shape, consisting of N 14-bit words, is loaded in to the FPGA via the serial link. On the rising edge of the TTL input, the N pulse-shape words are played out sequentially to the VGA DAC at 200 MHz. The final word of the pulse-shape is held on the DAC output. On the falling edge of the TTL input, the N pulse-shape words are played out backwards. Thus to generate a pulse with a $1\mu s$ rising and falling shape length we load a pulse-shape consisting of 200 words, with the first word being 0 (full-scale low) and the last being $2^{14} - 1$ (full-scale high).

B.2   Trap DC DACs

The trap DC voltages are supplied by a 5 channel DAC. Each channel can be set between $-240\,V$ and $+240\,V$ with a resolution of 7 mV. The DAC is programmed over an isolated serial link, connected to the experimental control computer by a USB-serial bridge.

The serial link, via an SPI isolator,[6] drives three $\pm 5\,V$ 16 bit dual DACs.[7] The DACs have integrated references with a 2 ppm/ K temperature coefficient. Each DAC output drives an output stage.

---

[6] Analog Devices ADUM1401.

[7] Analog Devices AD5752R.

The output stages (Fig. B.3) consist of a push-pull circuit driven by an opamp, with the opamp feeding back onto the divided output voltage. The transistor Q_1 acts as a level shifter for the opamp output, mapping the current through R_3 to the current through R_4, and hence V_{gs} of Q_3. The R_5, Q_2 and R_7, Q_5 pairs current limit the pull-up and pull-down current to $V_{BE}/R \sim 0.5$ mA. The diode D_1 makes the output stage push-pull: when Q_3 conducts to lower the output voltage, this diode drains the output capacitance. The output voltage is filtered by C_2, R_{10}, C_3, L_1, C_4. The opamp feedback loop is closed by the potential divider R_8, R_9 (48:1). This network dissipates up to 50 mW, and is built out of 10 ppm/ K resistors, hence is likely the limiting factor in the output voltage stability. With the circuit thermally lagged, we measure an output stability of 0.6 mV over 2 h (2 ppm), comparable with the measurement accuracy. The measured output noise is $\approx$1 mV rms in 0−20 MHz.

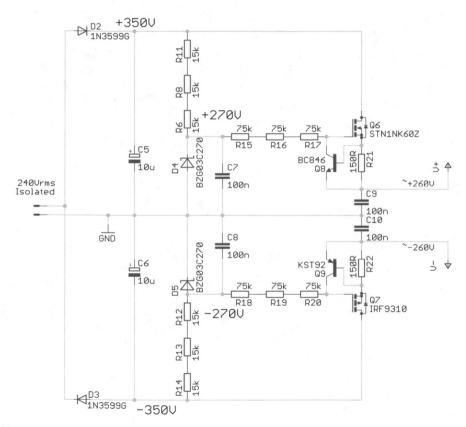

Fig. B.4 Trap DC DACs high-voltage power supply. The input comes from a 1:1 isolation transformer. These current-limited regulated supplies drive the 5 output stages (Fig. B.4)

The output stages are powered by an isolated high-voltage power supply (Fig. B.4). The 240 V AC input from an isolation transformer is half-wave rectified to give ± 350 V. A Zener diode regulator with a current limiter (Q_8, R_{21} and Q_9, R_{22}) produces a stable output of ± 260 V.

Appendix C
Readout Error Normalisation

In this appendix we discuss how we measure and normalise out the readout errors for one and two ions, and how the readout errors affect our measurement of Bell state fidelity.

C.1 Readout of a Single Ion

The readout process (state-selective shelving and fluorescence detection) acts as a linear map between the spin-state and the state determined from the readout

$$\begin{pmatrix} P_f \\ P_d \end{pmatrix} = \begin{pmatrix} 1 - \epsilon_\uparrow & \epsilon_\downarrow \\ \epsilon_\uparrow & 1 - \epsilon_\downarrow \end{pmatrix} \begin{pmatrix} P_\uparrow \\ P_\downarrow \end{pmatrix} \tag{C.1}$$

where $\epsilon_\uparrow$ and $\epsilon_\downarrow$ are the readout errors, P_f is the probability of measuring the ion to be fluorescing, and $P_d = 1 - P_f$ is the probability of measuring the ion to be 'dark' (not fluorescing). If the readout errors are non-zero we can still determine the spin-state accurately by applying the inverse of Eq. C.1 (having measured the readout errors in a separate experiment).

C.2 Readout of Two Ions

If we can distinguish the fluorescence from each ion, for example with a CCD, the two ion case is the same as the single ion case. However, if we detect just the total fluorescence, for example with a PMT, we can only resolve the number of ions fluorescing. Assuming the readout errors are identical we find

© Springer International Publishing AG 2017
C.J. Ballance, *High-Fidelity Quantum Logic in Ca+*,
Springer Theses, DOI 10.1007/978-3-319-68216-7

$$
\begin{pmatrix} P_{ff} \\ P_{df} + P_{fd} \\ P_{dd} \end{pmatrix} = \begin{pmatrix} (1-\epsilon_\uparrow)^2 & \epsilon_\downarrow(1-\epsilon_\uparrow) & \epsilon_\downarrow^2 \\ 2\epsilon_\uparrow(1-\epsilon_\uparrow) & (1-\epsilon_\uparrow)-\epsilon_\downarrow(1-2\epsilon_\uparrow) & 2\epsilon_\downarrow(1-\epsilon_\downarrow) \\ \epsilon_\uparrow^2 & \epsilon_\uparrow(1-\epsilon_\downarrow) & (1-\epsilon_\downarrow)^2 \end{pmatrix} \begin{pmatrix} P_{\uparrow\uparrow} \\ P_{\uparrow\downarrow} + P_{\downarrow\uparrow} \\ P_{\downarrow\downarrow} \end{pmatrix}
$$

$$(C.2)$$

Inverting this, and writing the result in terms of the 'two ion bright' probability, P_{ff}, and the 'one ion bright' probability, $P_{1f} = P_{df} + P_{fd}$

$$
\begin{pmatrix} P_{\uparrow\uparrow} \\ P_{\uparrow\downarrow} + P_{\downarrow\uparrow} \\ P_{\downarrow\downarrow} \end{pmatrix} = \frac{1}{(1-\epsilon_\uparrow-\epsilon_\downarrow)^2} \begin{pmatrix} \epsilon_\downarrow^2 - \epsilon_\downarrow P_{1f} + (1-2\epsilon_\downarrow)P_{ff} \\ -2\epsilon_\downarrow(1-\epsilon_\uparrow) + (1+\epsilon_\downarrow-\epsilon_\uparrow)P_{1f} + 2(\epsilon_\downarrow-\epsilon_\uparrow)P_{ff} \\ (1-\epsilon_\uparrow)^2 - (1-\epsilon_\uparrow)P_{1f} - (1-2\epsilon_\uparrow)P_{ff} \end{pmatrix}
$$

$$(C.3)$$

We can determine $\epsilon_\uparrow$ and $\epsilon_\downarrow$ by preparing $|\downarrow\downarrow\rangle$ and $|\uparrow\uparrow\rangle$ and reading-out

$$|\uparrow\uparrow\rangle : P_{ff} = (1-\epsilon_\uparrow)^2, \quad P_{1f} = 2(1-\epsilon_\uparrow)\epsilon_\uparrow \qquad (C.4)$$

$$|\downarrow\downarrow\rangle : P_{ff} = \epsilon_\downarrow^2, \quad P_{1f} = 2(1-\epsilon_\downarrow)\epsilon_\downarrow \qquad (C.5)$$

As we expect $\epsilon_\uparrow, \epsilon_\downarrow \ll 1$ (that is, our readout to be pretty good) the 'one ion bright' signal is the most sensitive to the readout errors. Solving for the readout errors, we find

$$\epsilon = \frac{1}{2}\left(1 - \sqrt{1-2P_{1f}}\right) \approx \frac{1}{2}P_{1f} \qquad (C.6)$$

where ϵ is the upper readout level, $\epsilon_\uparrow$, if we prepared $|\uparrow\uparrow\rangle$, or $\epsilon_\downarrow$, if we prepared $|\downarrow\downarrow\rangle$. The approximation is for small P_{1f}.

C.3 Effect of Shelf Decay

In our treatment so far we have neglected any decay of our shelf state. When reading out a single ion the error from shelf decay is indistinguishable from any shelving error, and hence is completely described by the linear map of Eq. C.1. However when we read out two ions this effect becomes significant. As the background count rate is much smaller than the ion fluorescence count rate the detection error from shelf decay depends on the total number of ions fluorescing. This means that the readout process linear map can no longer be expressed as a tensor product of two single-ion readout processes (as we assumed in Sect. C.2).

We can estimate how significant this error is by simulating a fluorescence detection experiment with shelf decay using our typical experimental background and fluorescence rates. We can then simulate the measurements we normally make to determine the readout levels, and then normalise them out using the linear map of Sect. C.2. We then can see how faithfully our naïve readout normalisation infers the correct spin state.

Performing this simulation we find that, as expected due to the symmetry, there is no systematic error is measuring $|\downarrow\downarrow\rangle$ or $|\uparrow\uparrow\rangle$, but there is a systematic underestimate of $|\uparrow\downarrow\rangle$ of $\approx 2 \times 10^{-4}$. Simulating a gate fidelity measurement we find that this leads to systematic fidelity underestimate of $\approx 1 \times 10^{-4}$. As these systematics are smaller than our typical statistical errors we ignore them in this thesis.

C.4   Fidelity of Bell States

We want to calculate how our measured Bell state fidelity changes with imperfect readout. We assume the spin-state is a perfect Bell state. We model a fidelity measurement, following the method described in Sect. 8.1, after propagating the spin-state through Eq. C.2. The fidelity error we measure is

$$
1 - \mathcal{F} = \bar{\epsilon} - \frac{1}{2}(\epsilon_\uparrow^2 + \epsilon_\downarrow^2)
$$
$$
+ 2\bar{\epsilon}(1 - \bar{\epsilon})
$$
$$
\approx 3\bar{\epsilon} \tag{C.7}
$$

where $\bar{\epsilon} = \frac{1}{2}(\epsilon_\uparrow + \epsilon_\downarrow)$. The first line is the contribution from the population, and the second from the parity contrast.

References

[Szw09]   Szwer, D.J. 2009. *High Fidelity Readout and Protection of a 43Ca+ Trapped Ion Qubit*. Ph.D thesis, University of Oxford.
[MvdS99]   Metcalf, H.J., and P. van der Straten. 1999. *Laser Cooling and Trapping*. Berlin: Springer.

Printed in the United States
By Bookmasters